Essentials of RF and Microwave Grounding

For a complete listing of recent titles in the *Artech House Microwave Library*, turn to the back of this book.

Essentials of RF and Microwave Grounding

Eric Holzman

ARTECH
HOUSE

BOSTON | LONDON
artechhouse.com

Library of Congress Cataloging-in-Publication Data
A catalog record for this book is available from the U.S. Library of Congress.

British Library Cataloguing in Publication Data
Holzman, Eric
    Essentials of RF and microwave grounding.—(Artech House microwave library)
    1. Electric currents—Grounding   2. Microwave transmission lines—Design and
    construction   3. Electric circuits—Design and construction
    I. Title
    621.3'17

    ISBN-10: 1-58053-941-6

Cover design by Igor Valdman

International Standard Book Number: 1-58053-941-6

10 9 8 7 6 5 4 3 2 1

To Ingrid

Contents

Preface

Grounding is about the flow of electrons on conductors. At low frequencies, a source of electrons moves them through a closed circuit comprised of conductors and loads such as resistors, inductors, capacitors, and transistors. The electrons return to the source's grounded cathode via the lowest impedance or ground path. The source reenergizes the electrons and sends them around the circuit once again. For those who work with low frequency electric circuits and equipment, proper grounding is synonymous with safety: keeping the electrons that flow through the equipment isolated from its users.

For RF and microwave engineers—the designers of high frequency electrical circuits and antennas—proper grounding means much more than safety. The rapidly alternating currents of microwave circuits also flow on conductors, but the circuits are no longer strictly closed. One circuit's currents may transmit energy to the currents on another, physically removed circuit. If the high frequency grounding of these circuits is not properly designed, they may malfunction and interfere with each other in unexpected and undesirable ways.

Microwave engineering focuses on the electromagnetic field, a nonphysical but very useful mathematical artifice, which describes the behavior of forces exerted over a distance between currents and charges. Many commonly used distributed circuit terms, such as impedance, are defined in terms of fields. In particular, antenna design is devoted almost completely to optimizing the radiation field and associated input impedance of a distributed structure. Our emphasis on this field-centric paradigm often causes us to forget that fields cannot exist without sources—that is, the electrons that are flowing on the conductors of microwave components. As with low frequency circuits, these electrons often flow from generators to loads, and the electrons must return via a low impedance path to ground. As frequency increases, however, the increasing inductive reactance inherent even in good ground conductors can seriously degrade the performance of a microwave circuit. Furthermore, because a microwave circuit's physical size is on the order of a wavelength, the length of the ground path matters also.

Engineers specializing in electromagnetic compatibility/electromagnetic interference (EMC/EMI) are well acquainted with these grounding issues, but this is not a book on shielding and the suppression of unwanted radiation from microwave equipment. The goal here is to describe techniques for RF grounding that should be used by the designers of microwave circuits, components, and antennas. These techniques will likely mitigate EMC/EMI problems, but just as importantly, they will enable high frequency electronic components to reach their maximum performance.

To appreciate the subtleties of high frequency grounding, one needs to understand clearly a number of fundamental concepts; so this text contains sufficient background material to be self-contained. Beyond the fundamentals, the beginning of each chapter is spent reviewing pertinent material before going on to explain the influence of grounding. Because of the breadth of the coverage—grounding is important for just about every microwave device—we use simple derivations and results from numerical electromagnetic simulations of real microwave components to focus the reader's attention on the path taken by the neglected flowing electrons. Performance problems that occur when grounding design is inadequate are highlighted along with methods to avoid them. Although the selection of topics must necessarily be limited, the coverage is sufficiently broad to enable the reader to acquire an intuitive and physical introduction to microwave grounding, one that can be used to solve problems not covered here.

The first two chapters provide relevant background material. Chapter 1 gives an overview of the topics covered in the book and introduces low frequency grounding, starting with simple lumped circuit examples. The low frequency definition of grounding is broadened to cover distributed circuits and antennas and is followed by a list of the key problems of poorly designed grounding paths for microwave circuits: breaks, excessive length, and high impedance.

Chapter 2 reviews electromagnetic theory, starting with Coulomb's postulate for the force between two charges and progressing quickly to steady state, time harmonic fields. Grounding is defined more precisely from the perspective of electromagnetics. The chapter concludes with an introduction to radiation.

Many grounding problems involve microwave transmission lines and components constructed from them, so Chapter 3 discusses a variety of conductor-based transmission lines, including coax, microstrip, and waveguide. The flow of currents on these transmission lines is examined. For single conductor circuits constructed from waveguide, the meaning of ground can be ambiguous, and thus the current path inadvertently may be physically cut, causing excessive loss and unexpected radiation. The coverage of planar transmission lines includes an in-depth discussion of RF grounding for multilayer, mixed signal printed circuit boards and surface mounted microwave components. The differences between DC and RF short circuits as they pertain to grounding at high frequencies are also presented, including a detailed look at via holes in the ground path.

Continuing with transmission lines, Chapter 4 discusses grounding and transitions between different types of transmission lines. Care in the design of the ground path is essential for optimum performance, as results from numerical electromagnetic simulations of transitions between microstrip, coax, and waveguide help to illustrate.

Chapter 5 looks at grounding in active microwave component design. First, we examine components constructed from diodes, including switches and mixers. A significant portion of the chapter is devoted to microwave field-effect transistors, which are the building blocks of most active microwave components in use today. Simple derivations and results from a microwave circuit simulator illustrate the consequences of poor source grounding. The design of multistage amplifier chains with grounded shields for the suppression of feedback oscillations is discussed also. The last section of the chapter covers grounding of active devices on printed circuit boards.

Antennas and ground planes are the topic of the Chapter 6. As with microwave circuits, the currents flowing on antennas are what matter. But unlike circuits, antennas are designed so these currents radiate as efficiently as possible. The control of this radiation can be a challenge, and it is not uncommon for a newly designed antenna, placed in its real-world environment, to exhibit a badly distorted radiation pattern, a problem that can be acute for broad-beamwidth antennas. In addition, the resonant frequency of a narrow band antenna such as a dipole often shifts when it resides near other ground planes, such as those found on printed circuit boards. Such problems arise because the currents flowing on the antenna interact with currents on other conductors in the vicinity, as a cell-phone antenna might interact with its user's body. A number of methods for controlling this interaction are described.

It is amazing and rewarding to watch a tangible thing such as a book be born from one's own thoughts. Along the way, I have been lucky to benefit from the thoughts and efforts of others. First, many of us take the availability of three-dimensional numerical electromagnetics software for granted, but I am thankful to have been able to use Computer Simulation Technology's Microwave Studio and Agilent's Advanced Design System to analyze many of the examples in Chapters 3 through 6. In addition, several of my coworkers at my former employer, Telaxis Communications, have contributed to this book.

Salvador Ramirez-Rivera provided the design data on the two waveguide-to-microstrip transitions used in Chapter 4, the subharmonic mixer example in Chapter 5, and he reviewed Chapter 5 with great care. Chris Koh graciously let me use portions of his memorandum on module grounding, also in Chapter 5. Kyle Watson provided the drawing appearing in Figure 6.40, and George Winslow provided the drawing appearing in Figure 3.39(b). Chris, Salvador, Kyle, and George worked with me at Telaxis under the leadership of Ken Wood. Ken was instrumental in establishing the creative environment under which many of the ideas and material in this book were conceived. In bringing this project to fruition, the support from the Artech House staff has been nothing less than professional. The technical reviewers' comments were detailed and thoughtful and certainly made this a better book. Barbara Lovenvirth, my developmental editor, helped smooth the bumps during the writing process. I greatly appreciate her assistance, particularly in obtaining the technical reviews. I also owe a debt to Mark Walsh for his enthusiasm and interest in this project, which motivated me to transform a one-page outline that had sat on my desk for a year into the book before you. Finally, during the past year, as I wrote this book, my wife, Ingrid, and my children, Dirk and Rya, dealt with the stress of adjusting to a new home. Even so, they remained patient, supportive, and loving, reminding me of the most important things in my life.

1

Introduction to Grounding

In this chapter, we review the fundamentals of low frequency grounding and define the following terms: ground, ground path, and grounding. We compare low frequency and high frequency circuits and describe the differences between low frequency and high frequency grounding. We conclude with a summary of radio frequency (RF) grounding problems and their impact on the performance of microwave components and subsystems.

1.1 Grounding for DC and Low-Frequency AC Circuits

The principles of low frequency grounding are well known, and they form the basis for our discussion of high frequency grounding. Figure 1.1 shows a simple circuit consisting of a voltage source that supplies electrical current to a load such as a light bulb. The electric current I_S is defined in terms of positive charges moving from the positive electrode of the source to ground. Since electrons are negatively charged, they flow in the opposite direction. The source does work and transfers energy to the current I_S. The current flows in metal conductors, such as wires, to the load. If the connecting wires are lossless (meaning they do not reduce the electron potential), and the load is matched properly to the source, all the current's energy is transferred to the load. The current emerges from the load at a potential of 0 volts and returns to the source along the ground path. Since charge can neither be created nor destroyed, the source and load currents must be the same. Ohm's law gives the current I_S as

$$I_S = V_S / Z_L \qquad (1.1)$$

$V = V_S$ I_S

Signal path

Source   Load

$+$

V_S Z_L

$-$

I_S

$V = 0$ Ground path

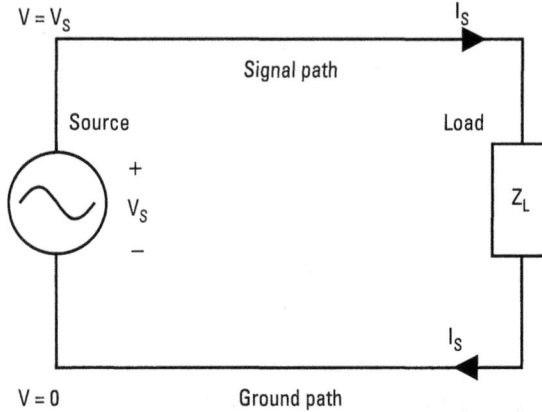

Figure 1.1 Idealized low frequency circuit showing signal and ground paths.

This low-frequency circuit truly is a closed loop in that it does not exchange energy via radiation with other circuits that may be nearby.[1] Aside from magnetic interactions, the signal and ground paths are isolated and may even be widely separated. By *signal path*, we mean the conductive path taken by the current flowing from the source; and by *ground path*, we mean the conductive path taken by the current returning to the source from the load. We connect the ground path to the negative terminal of the source, which sits at 0 volts potential. This choice of reference is arbitrary, since ground might just as well be connected to the source's positive terminal.

The National Electrical Code (NEC) defines ground as "a conducting connection, whether intentional or accidental, between an electrical circuit or equipment and the earth or to some conducting body that serves in place of the earth" [1]. In Figure 1.1, the negative terminal of the source serves in place of the earth. This terminal is the circuit's *ground.* An alternate definition by Ott is more precise [2]: "an equi-potential point or plane, which is a source or sink for current, and serves as a reference potential for a circuit or system." If the source in Figure 1.1 is an alternating current (AC) source, the voltage V_S at the anode will alternate between positive (source) and negative (sink) relative to $V = 0$, and the direction of current flow will alternate also. Even so, the voltage at ground stays at 0 volts, so the location of ground and the ground path do not change.

It follows that a *grounded* circuit is "*connected* to earth or to some conducting body that serves in place of the earth" [1]. Although ground is a point in a circuit, *grounding* involves the ground, the ground path, and being grounded.

---

1. A low frequency AC circuit can be coupled magnetically to another, as occurs in a transformer; and as we discuss in Chapter 2, parallel wires carrying DC current can interact magnetically.

In the case of microwave circuits, what matters most are the precise characteristics of the path to ground taken by the current returning to the source, and so we will focus in this book on a second definition of ground by Ott [2]: "a *low impedance* path for current to return to the source." Low impedance implies that there is a voltage difference between separated points along the ground path, as shown schematically in Figure 1.2. Since all conductors have resistance and inductance, the ground path has an impedance Z_G equal to $R_G + j\omega L_G$, where $\omega = 2\pi f$, and f is the frequency of the AC source. At or near 0 Hz the resistance is dominant, so increased current flow through the ground path causes an increased voltage drop. With increasing frequency, the conductor's reactance becomes dominant. Because inductance increases with conductor length, minimizing the length of the ground path becomes essential as frequency increases.

For AC circuits, the impedance in the ground plane combines with the load to determine the current flowing in the entire circuit. Now the source current is

$$I_S(\omega) = V_S / [Z_L(\omega) + Z_G(\omega)] = V_S / (Z_L(\omega) + R_G + j\omega L_G) \quad (1.2)$$

where we have assumed that the load impedance has a frequency dependence also. If the load is capacitive, it even can resonate with the ground plane's inductance at a particular frequency. In the case of a resonantly matched antenna such as a microstrip patch or dipole, reactance in the ground path can shift the resonant frequency or degrade the antenna's input match.

Most practical circuits have multiple, parallel ground paths, as shown in Figure 1.3. In this circuit, the source drives two parallel loads, Z_{L1} and Z_{L2}. The return current I_1 from load Z_{L1} must overcome ground path impedance Z_{G1},

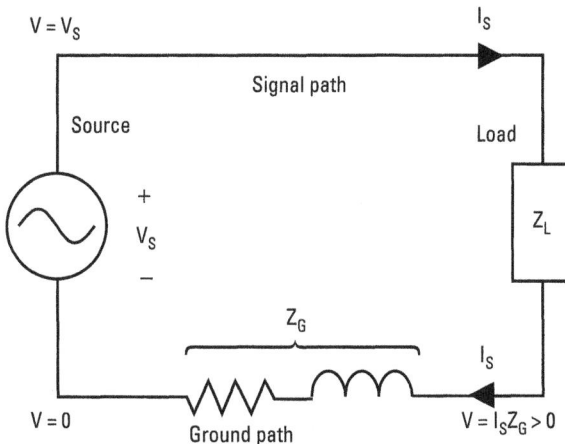

Figure 1.2 Circuit with impedance Z_G in ground path.

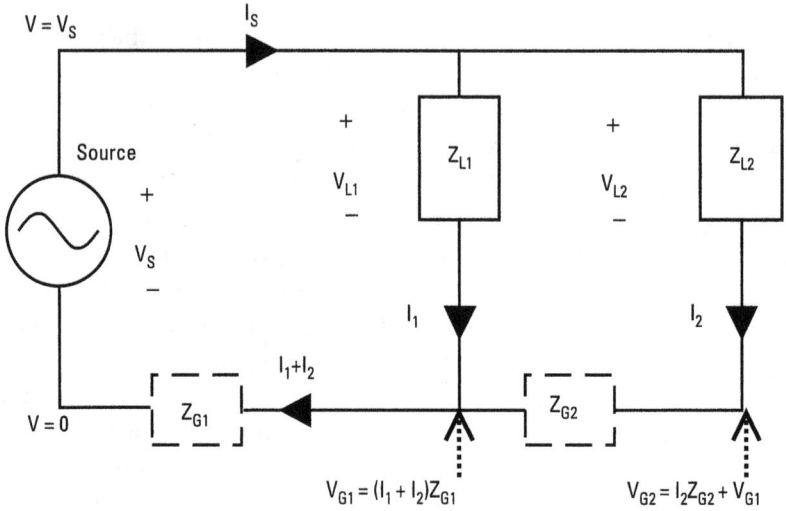

Figure 1.3 Multipoint grounding. Parallel loads Z_{L1} and Z_{L2} are coupled by impedance in the ground path. (*After:* [3].)

while the current I_2 returning from load Z_{L2} has further to travel, and must surmount impedance Z_{G2} in addition to Z_{G1}. The total current flowing from the source is

$$I_S = I_1 + I_2 \tag{1.3}$$

where we have suppressed the frequency dependence. From Figure 1.3, we see that an expression for I_1 can be determined from the voltage across load Z_{L1}:

$$I_1 = (V_S - V_{G1})/Z_{L1} = V_S/Z_{L1} - (I_1 + I_2)Z_{G1}/Z_{L1}$$

which we solve for I_1 to get a fundamentally important result:

$$I_1 = (V_S - I_2 Z_{G1})/(Z_{L1} + Z_{G1}) \tag{1.4}$$

For I_2 we can derive a similar equation or just use (1.3). Equation (1.4) shows that a ground plane having a nonzero impedance couples I_1 and I_2: current I_1 is a function of current I_2. Consequently, the voltage, V_{L1}, across load 1 is a function of both its current I_1, and the current flowing through load 2, I_2. To minimize this dependence, we must minimize the ground plane impedance, Z_{G1}. For electronic circuits, it often happens that there are multiple, parallel paths to ground, as shown in Figure 1.4. Within the ground plane a portion of each component's ground path is shared with the other components. Any impedance in the ground plane will couple the currents of the components on

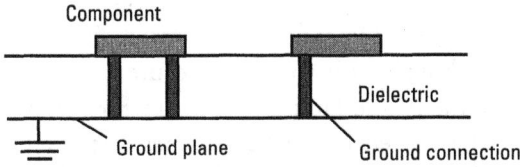

Figure 1.4 A circuit board has multiple connections to its ground plane. The ground plane is the shared portion of each component's ground path.

the circuit board. If one device is a noisy load such as a power supply, and the other is a sensitive detector diode, the ground plane's impedance will couple some of the noisy current flowing through the power supply to the detector and cause it to lose sensitivity.

For low frequency circuits, a single-point grounding scheme, such as the one shown in Figure 1.5, prevents ground currents from coupling. The ground paths leading away from the loads are completely isolated and connected only at a single ground point. Consequently, currents I_1 and I_2 are not interdependent despite the presence of impedance in their respective ground paths. For microwave systems, single-point grounding rarely is achievable without unacceptably long, and thus highly inductive, ground paths. Moreover, even if the conductors in a microwave circuit can be separated physically as in Figure 1.5, the RF currents can still couple through radiation.

There are a number of reasons why DC grounding is important, but the primary one is safety. For example, consumer electronic equipment and

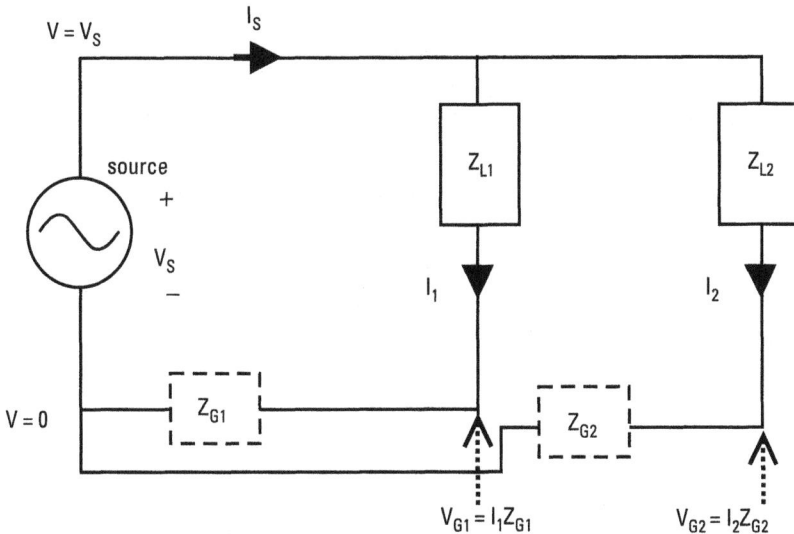

Figure 1.5 Single-point grounding. Ground paths are connected only at the ground point, so load currents do not couple.

appliances often are packaged in metal enclosures. A circuit fault that causes the internal circuitry to short circuit to the enclosure may develop. The current flowing in the equipment, the fault current, can now flow on the enclosure. When a person touches the enclosure, he becomes part of a multipoint grounding scheme: some of the fault current will flow through the enclosure and some through the person's body. Since the impedance of a person's body is high while the impedance of a well-grounded enclosure is extremely low, negligible current should flow through the person. However, if the enclosure is not grounded or is poorly grounded, a dangerously high portion of the fault current may flow to ground through the person's body. As an additional safety measure, a properly designed DC grounding circuit will include a protective device such as a circuit breaker to limit the amount of current that can flow if a fault does occur.

For most circuits, ground also provides the signal reference for the circuitry. Consequently, the effectiveness of DC and low frequency grounding depends on the path to ground, which should be designed intentionally and precisely known. The ground path must be permanent, electrically continuous, and capable of conducting ground current safely. A well functioning ground path maintains a low voltage between the load and ground, even for large currents, which facilitates operation of circuit protection devices and drains leakage, static, and unwanted noise-making currents to ground [4].

1.2   RF Grounding

The microwave circuit shown in Figure 1.6 illustrates some of the similarities and differences between RF and low frequency circuits.[2] Much like a low frequency circuit, this microwave circuit is driven by a source, and it has a signal conductor, the microstrip line, and a ground plane. The signal and ground current vectors are identical in magnitude, they have opposite polarity, and they will alternate in direction at the frequency of the source. The patch antenna, the load in this circuit, dissipates RF power through the phenomenon of radiation, a topic we will discuss in more detail in Chapters 2 and 6.

For most DC circuits, a schematic is sufficient to describe completely the electrical behavior of the circuit. The actual, physical construction of the circuit adds little to this description. On the other hand, most RF circuits cannot be described completely without a physical layout as shown in Figure 1.6. For DC circuits, the distance separating the signal and ground path generally is unimportant. In contrast, for an RF circuit, the separation of the signal and ground

---

2. Although RF can denote a specific band of frequencies in the electromagnetic spectrum, which is distinct from the microwave band, we will use the terms RF and microwave interchangeably throughout the text—a common practice.

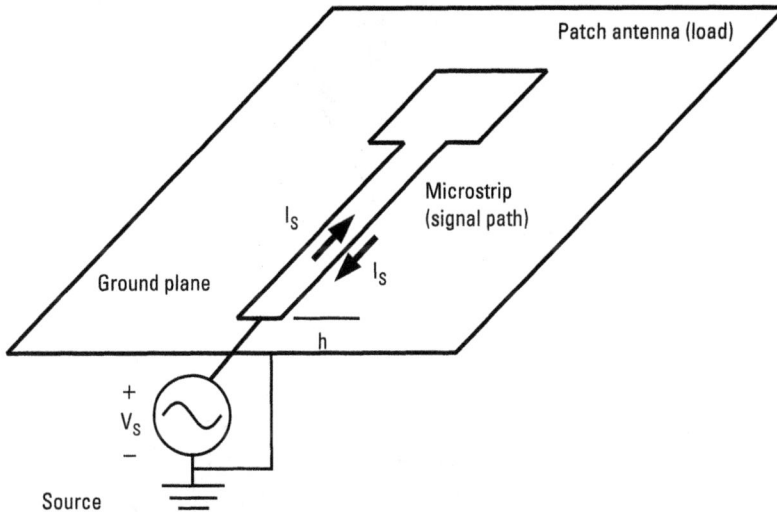

Figure 1.6 Microwave printed circuit showing source current flowing on microstrip and ground plane.

conductors establishes the configuration of the electric and magnetic fields along with their relationship, a ratio called the *wave impedance.* The dimensions of the patch antenna conductor and its distance from the ground plane determine the frequency at which its input impedance is resonant (pure real-valued). The ground plane provides the return path for current, so any impedance it possesses will add to the impedance of the antenna, and cause a change in its resonant behavior. The antenna match to the transmission line will degrade, and energy will be lost due to the mismatch and from dissipation in the ground path. Since the patch antenna, and even the microstrip line, can radiate, we cannot assume that this circuit is isolated from other RF circuits, such as those that may share the same circuit board.

We can extend our low frequency definition of ground to define an *RF ground* as a low impedance, nonradiating path to earth or a conducting body that serves in its place. Sometimes the same ground path may have to support DC and RF currents. The ground path's electrical characteristics will likely differ significantly at DC and RF frequencies, so we cannot assume that a good DC ground path is necessarily a good RF ground path.

As shown in Figure 1.6, the structure connecting the patch antenna with its source is a microstrip transmission line, a pair of conductors comprising a signal and ground that can transfer energy stored in the electromagnetic field from the source to the load. Figure 1.7 shows two other guided wave structures, a three-wire transmission line and a rectangular waveguide. The two outer conductors of the three-wire line commonly provide the ground path, with each carrying half the ground current, while the middle conductor carriers the full

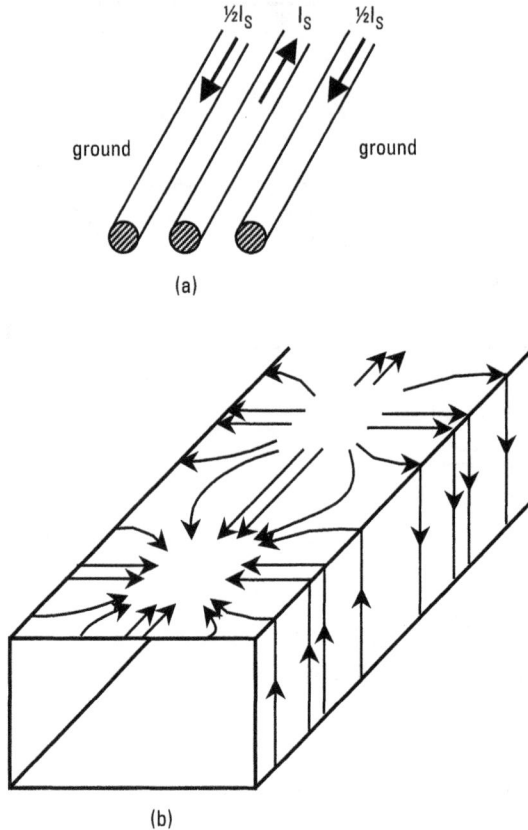

Figure 1.7 Where the current flows in a microwave structure depends on its geometry: (a) three-wire transmission line; and (b) dominant mode in rectangular waveguide.

signal current. On the other hand, the rectangular waveguide is constructed from a single conductor, and does not have physically distinct signal and ground conductors. The same conductor acts as the source and sink for RF current. We will discuss the grounding of transmission lines and waveguide components in Chapters 3 and 4.

A *passive* microwave circuit, such as an antenna or transmission line, does not require a DC voltage or bias to operate. Passive circuits only direct the flow of electrons from a source to a load. The source of the electrons is an *active* microwave component, such as the transistor amplifier that is shown schematically in Figure 1.8. The amplifier requires DC bias to operate and it outputs an RF current, so both DC and RF ground paths are required. From the schematic, we can see that the DC and RF ground paths share portions of the same conductor. While the schematic is sufficient to describe the DC grounding, we would need to see a layout to determine whether the RF grounding is adequate. For example, if this amplifier resides on a printed circuit board, we must establish

DC bias path

RF output

DC & RF input

V_{DS}

$+$
V_{RF}
$-$

RF
FET

Z_L

DC ground path

RF ground path

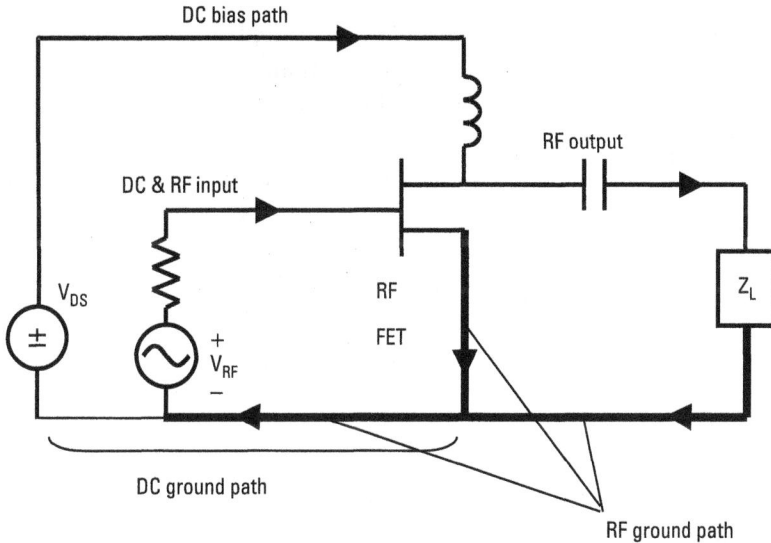

Figure 1.8 Microwave transistor amplifier schematic showing ground paths for DC and RF currents.

whether the amount of inductance in the field effect transistor (FET)'s RF ground path is small enough for stable operation by measuring the dimensions (length, width, and thickness) and computing the inductance of the ground path through the circuit board. In Chapter 3, we discuss printed circuit board grounding, and in Chapter 5, we discuss active microwave circuit design and grounding.

1.3   Why RF Grounding Is Important

Simply put, good RF grounding is an essential part of a well-designed microwave circuit. For passive microwave circuits, such as filters, poor grounding means high insertion loss, poorly matched input ports, mediocre isolation between input and output, and unwanted radiation. The bandwidth of transmission line transitions narrows when the ground path is not carefully designed. A microwave FET amplifier having excessive inductance in its source-to-ground path may oscillate and even fail. Similarly, a chain of amplifiers possessing high power gain requires careful attention to grounding to suppress output-to-input feedback and prevent oscillation. Diode-based switches require good grounding to achieve the highest input-to-output isolation. Poor DC grounding can cause voltage-controlled oscillators (VCOs) to have poor frequency stability and phase-locked loops (PLLs) to have high phase noise. For an antenna, insufficient grounding can mean poorly controlled radiation and the deterioration of its input impedance match. Subsystems, such as printed circuit boards, can experience all these problems when their grounding is inadequate.

There are a number of ways to insure good RF grounding during the design stage. In particular, careful, detailed modeling of the ground path is crucial. Towards this end, commercially available generalized electromagnetic analysis software has become so fast, accurate, and ubiquitous that microwave engineers frequently abandon the empirical trial and error design techniques of the past, particularly in the initial design stage. This reliance on simulation requires the designer to exercise caution, because circuits and components designed and simulated on the computer often are idealized versions of the actual hardware that is built and tested. Discontinuities are left out; gaps are omitted; conductivity is infinite; and current path lengths are shortened unintentionally. All these idealizations can lead to grounding problems when hardware is built. A microwave engineer who knows how to design a RF ground path is more likely to achieve success with his first experimental prototype.

Throughout this book, we will examine how a group of basic RF grounding problems influences the performance of microwave circuits and antennas. These problems include: (1) discontinuities, (2) excessive inductive reactance, (3) excessive resistance, (4) excessive electrical length, (5) incorrect ground-signal conductor geometric relationships, (6) ground path induced coupling, and (7) in the case of antennas, carelessly configured or located ground planes. We recommend solutions based on the use of correct RF grounding techniques that must be applied during the *design* of microwave components.

Performance problems inevitably arise when an RF component, subsystem, or system has grounding problems. For simple components, these problems can usually be isolated relatively quickly, but a redesign is often necessary. For more complex microwave systems such as wireless transceivers, the diagnosis of grounding problems can take significant time and effort, and a careful redesign of the system is almost certainly required. Hopefully, the redesigned system does not have new problems. The conclusion is obvious: make every effort to prevent grounding problems before your design leaves the drawing board.

References

[1]   "The National Electrical Code," Quincy, MA: National Fire Protection Association, NFPA 70-1996, 1995.

[2]   Ott, H. W., "Ground—A Path for Current Flow," *Proceedings of IEEE International Symposium on Electromagnetic Compatibility*, 1979, pp. 167–170.

[3]   Paul, C. R., *Introduction to Electromagnetic Compatibility*, New York: Wiley Interscience, 1992, p. 701.

[4]   O'Riley, R. P., *Electrical Grounding*, 4th ed., Albany, NY: Delmar Publishers, 1996, p. 11.

2

Electromagnetic Theory

Microwave engineers design RF circuits and antennas with solutions to Maxwell's equations, which describe mathematically the interaction at a distance between electromagnetic sources (charges and currents) and materials (dielectrics, semiconductors, and conductors) in space and time. Our study of grounding concentrates on the flow of current in a variety of conductors such as wires, printed circuit boards, and antennas. Maxwell's equations help us to understand precisely the behavior of ground currents in these various conducting structures. In this chapter, we review pertinent aspects of electrostatics, magnetostatics, and electromagnetics and solidify the concepts of grounding introduced in Chapter 1.

2.1 Microwave Engineering—Focus on the Electromagnetic Field

The field of electrical engineering is vast and necessarily divided into a large number of specialties that are mostly application related. A unifying perspective of electrical engineering focuses on the frequency of the signals that are processed by an electrical component such as DC and low frequency AC (to a few kilohertz), high frequency (3 to 30 MHz), microwave (1 to 30 GHz), and millimeter wave (30 to 300 GHz). In each of these portions of the electromagnetic spectrum, the relationship between the wavelength (light speed divided by frequency) of the electromagnetic signal and the circuit size dictates the necessary design approach. At low frequencies, the dimensions of the circuit are minute compared to a wavelength, so we can ignore the propagating phase delay of the signal and design devices using the rules of lumped circuits. No precisely defined boundary between low and high frequency circuit design exists. High-speed digital circuits operate above 1 GHz, and large microwave systems can operate at frequencies in the hundreds of megahertz. Essentially, we are in the realm of microwave engineering when the dimensions of our circuit are comparable to a wavelength. Such a *distributed* circuit, antenna, or system takes advantage of the electromagnetic energy that can be transferred by and between physically separated, time-varying currents. Microwave design involves the careful selection of the dimensions of a distributed device and the material it comprises to influence the phase of the electrical signal as it propagates through the device at the desired frequency.

Since Faraday and Maxwell's pioneering work in the later half of the nineteenth century, the electromagnetic field, which describes the distribution of the space-time forces that arise from charges and currents, has been the primary tool of microwave engineering. The field-based approach to electromagnetic analysis and design, a purely mathematical representation of a spatial interaction, has proven to be accurate, efficient, and thus dominant. Although field theory certainly accounts for the sources of the field, currents, and charges, these are not its focus.

Before Faraday and Maxwell, scientists did not think in terms of an electromagnetic field, but viewed charges and currents as centers of force acting on each other at a distance. In Maxwell's words, "Faraday saw a medium where they saw nothing but distance..." [1]. We are not about to return to this source-centric paradigm, but since proper grounding technique involves the visualization of the currents that flow in a circuit, we will study more closely the sources of the electromagnetic field in the discussion that follows.

2.2 Electrostatics and DC Ground

Electrostatics is a special case of electromagnetic theory for which the electric field is static, and the only sources are charges at rest. It provides the theoretical foundation for the concept of DC ground.

2.2.1 Coulomb's Postulate and the Electrostatic Field

All of electrostatic theory is based on Coulomb's postulate, which says that the *force* between two stationary, charged particles is proportional to the product of the charges and inversely proportional to the square of their separation. The postulate is illustrated by way of the diagrams in Figure 2.1. The force exerted by particle 2 on particle 1 is given by

$$F_{12} \sim \mathbf{1}_{R12}\, q_1 q_2 / R_{12}^{\;2} \tag{2.1}$$

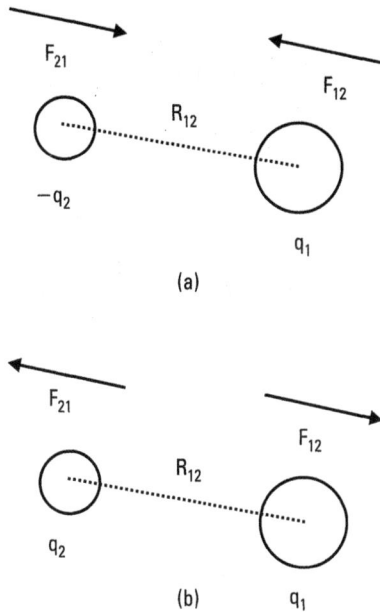

Figure 2.1 Forces on two fixed charges in space: (a) opposite polarity attract; and (b) same polarity repel.

where q_1 and q_2 are the charges on the particles, R_{12} is the distance between their centers, and 1_{R12} is the unit vector pointing in the direction of the force exerted on particle 1 by particle 2.[1] Figure 2.1(a) shows oppositely charged particles, such as protons and electrons, under an attractive force; while Figure 2.1(b) shows the repulsion of particles having the same charge polarity.

The *electric field intensity* E is defined as the *force* per unit charge that a test charge experiences when it is placed in a region where it is subject to interaction with other charged particles. E is a vector directed parallel to F, defined by F = qE, where q is the test charge. The units of E are newtons/coulomb = volts/meter. Although a single charged particle has an electric field, the particle does not exert an electrostatic force on itself. Thus, within a region of empty space, no force is exerted without the presence of at least *two* charged particles. The electric field of a single particle describes the *potential* for interaction should another particle be brought within a finite distance. For example, if two charged particles are held by some means fixed in space as in Figure 2.1(b), they exert forces on each other that make them want to move apart much like a spring under compression. Like the spring, this two-particle system stores potential energy.

---

1. Throughout this book, vector quantities such as E are printed in boldface type to denote their directional nature.

Given Coulomb's postulate, we can show that two equations describe the characteristics of the electrostatic field, namely,

$$\nabla \cdot \mathbf{E} = \rho / \varepsilon_0 \qquad (2.2)$$

$$\nabla \times \mathbf{E} = 0 \qquad (2.3)$$

where ρ is the electric volume charge density in coulombs/meter^3, and ε_0 is the permittivity of free space, 8.854×10^{-12} farads/meter. ∇ is the vector del operator, given by

$$\nabla = \mathbf{1}_x \, \partial / \partial x + \mathbf{1}_y \, \partial / \partial y + \mathbf{1}_z \, \partial / \partial z$$

in Cartesian coordinates, where $\mathbf{1}_x$, $\mathbf{1}_y$, and $\mathbf{1}_z$ are unit vectors in the x, y, and z directions. Equation (2.2) states that if a net electric flux leaves a volume, there must be positive charge inside. Equation (2.3), which says the electrostatic field is not rotational, has the solution

$$\mathbf{E} = -\nabla \Phi \qquad (2.4)$$

where Φ is the *electrostatic potential* in volts of a positive unit charge placed at a point (x,y,z). For an arbitrary volume charge distribution, $\rho(x', y', z')$, the potential is given by

$$\Phi(x, y, z) = \int_v \rho(x', y', z') \, dx' dy' dz' / 4\pi\varepsilon_0 (r - r') \qquad (2.5)$$

The potential due to a surface charge distribution has a similar equation with a surface integral replacing the volume integral. Equi-potential lines and surfaces are perpendicular to the direction of the electric field. Figure 2.2 shows a single, positively charged particle and a test charge used to probe its electric field. The electric field lines are directed uniformly and radially away in all directions from the particle towards infinity. Φ_1 and Φ_2 are equi-potential surfaces with $\Phi_2 > \Phi_1$. We can measure the electric potential directly and infer $\mathbf{E}$, but we cannot measure $\mathbf{E}$ directly.

Work is done in the field when we move a charge in the direction against it. In so doing, we increase the electrostatic potential of the charge and add potential energy to the system. We define $|\mathbf{E}|^2$ as power, and the energy stored in the electrostatic field is

$$W = (\varepsilon_0 / 2) \int_v |\mathbf{E}|^2 \, dx \, dy \, dz \qquad (2.6)$$

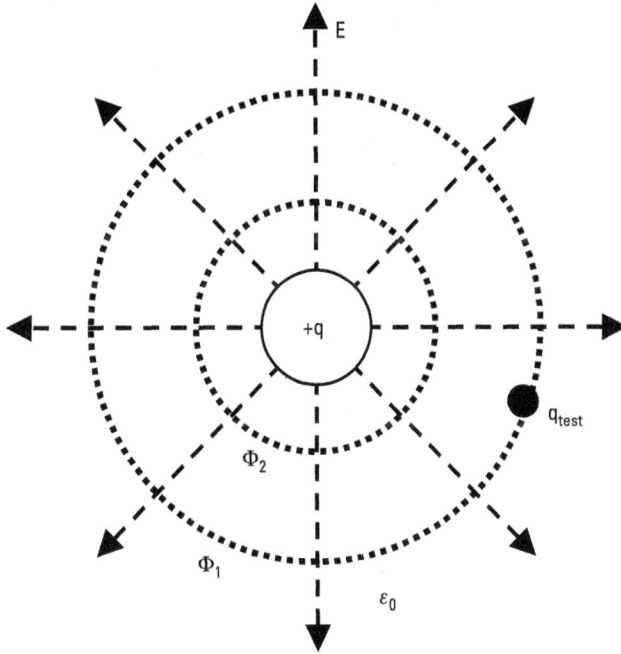

Figure 2.2 Charge and small test charge used to probe its electric field.

Physically, this energy must be associated with the charge configuration of the system, but field theory takes the view that it is stored in the field [2].

2.2.2 Conductors

Grounding is about the flow of charges in or on conductors. A conductor is a type of material. Materials can be described in microscopic terms as made of atoms, each of which consists of a positively charged nucleus having electrons orbiting in a number of discrete energy bands. An electron residing in the outer-most band, the *valence band,* may break free from the nucleus if it has sufficient energy. In conductors the electrons in the valence band are loosely bound and can migrate easily between atoms. In metals, there are many such free electrons. Above absolute zero, these electrons will move in random directions throughout the metal as in Figure 2.3(a) so that the net flow of current is zero. Conse-quently, if we place a group of charges inside a conductor, the free electrons in the conductor will exert a force on the group, which will force them apart and towards the surface of the conductor. The electrons will rapidly redistribute themselves so that there is no net charge inside the conductor. Because the net charge inside the conductor is zero, (2.2) says the electric field inside must also be zero.

On the surface of a conductor, the tangential electric field must be zero:

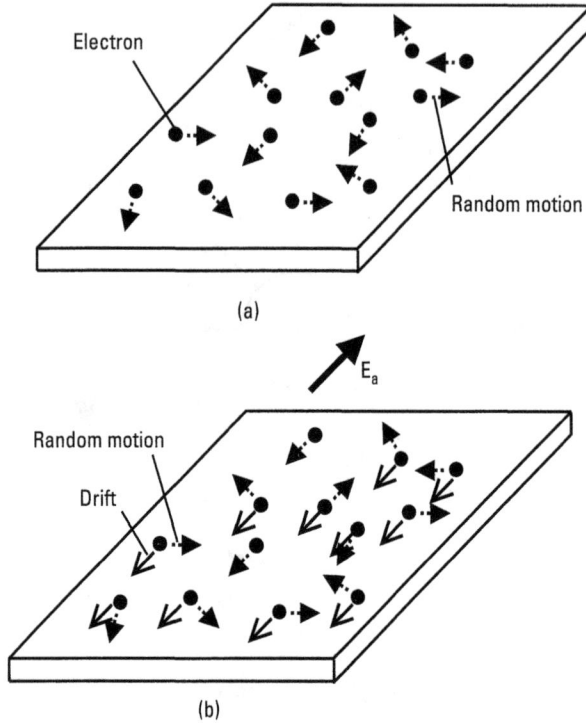

Figure 2.3 (a) Free electrons on a conductor move randomly. (b) When an electric field is applied, they drift in direction opposite to that of applied field.

$$E_{tan} = 0 \qquad \text{on the surface of a conductor} \qquad (2.7)$$

because any field that existed would cause free electrons to move very rapidly in such directions as to counteract it. Consequently, under equilibrium, the surface of a conductor is an equi-potential surface:

$$\Phi = \text{constant} \qquad \text{on the surface of a conductor} \qquad (2.8)$$

In a good conductor like copper, the time it takes charges to redistribute themselves is of the order of 10^{-19} seconds [3].

Additionally, since electric flux lines are terminated normal to the surface of conductors on charges, we can write the following boundary condition for the normal electrostatic flux density on a conductor:

$$D_n = \varepsilon E_n = \rho_s \qquad \text{on the surface of a conductor} \qquad (2.9)$$

where D_n is the normal component of the electric flux density in coulombs/meter^2, and ρ_s is the surface charge density on the conductor.

When an electrostatic field is applied across a conductor, superimposed upon the free electron's random motion is a net *drift velocity* in the direction opposite to that of the applied field as shown in Figure 2.3(b). The drift velocity is defined in terms of the applied electric field by

$$\mathbf{v} = \mu \mathbf{E} \qquad (2.10)$$

where μ is the *mobility* of the drifting particles. The mobility is a function of the material properties and the magnitude of the applied field. Since charges are moving, we have conduction current. The cross-sectional current density in amps/meter^2 is given by

$$\mathbf{J} = q\mathbf{v} \qquad (2.11)$$

where q is the volume charge density. We can define the *static conductivity* from

$$\mathbf{J} = \sigma \mathbf{E} \qquad (2.12)$$

where the conductivity is given by

$$\sigma = q\mathbf{v}/\mathbf{E} = q\mu \qquad (2.13)$$

A material is classified as a conductor, semiconductor, or insulator, depending on the value of its conductivity.

Conductivity is inversely related to resistivity, ρ, and the resistance of a piece of material of uniform cross-sectional area A and length L is $R = L\rho/A$. Thus, increasing the cross-sectional area of a conductor in the ground path will lower its resistance.

An electric *short circuit* occurs when an electric conductor joins two isolated charge distributions and provides a zero resistance path between the charges. The charges will redistribute themselves to eliminate any unbalanced electric fields. The short circuit is the path of least resistance between the original two charge distributions, so if the charges are of opposite polarity, the charges at the higher potential will flow through the short circuit towards the lower potential charges. The conductor potential will become uniform as the charges redistribute themselves over its surface.

2.2.3 Semiconductors and Dielectrics

Besides conductors, there are two other classes of electrical materials. *Insulators* or *dielectrics* have very low conductivity because their valence electrons are bound so tightly to the nucleus that they can only be freed to conduct by very strong applied forces (fields). In a perfect insulator, $\sigma = 0$, and conduction current cannot flow, no matter how large the electric field magnitude might be. But every material has some conductivity. When sufficiently large charges of opposite polarity accumulate across an insulator, it can break down—conduct electrons or arc. *Electrostatic discharge* (ESD) occurs when charge accumulates on one object, such as a person's body, with sufficient potential to arc through the air to another object, such as a sensitive electronic device. Proper low frequency grounding can prevent the ESD damage that can result. Wrist straps made from conductive materials commonly are used to conduct to ground the charge that builds up on the bodies of technicians who assemble and repair sensitive electronic equipment. On a much larger scale, lightning is ESD in the atmosphere between charges in the clouds and on the ground. Like any current, the lightning follows the least resistive path to ground, which often means it take the shortest path, touching down at high points on the Earth's surface. A lightning rod is an intentionally designed ground path for lightning, its purpose being to conduct lightning down a safe path to Earth ground. A properly designed lightning rod provides the lowest resistance path to ground in the vicinity of its installation.

The third type of material, a *semiconductor*, possesses some loosely bound charges in the valence band, although the quantity is small compared to that of conductors. The semiconductor *pn junction* is the building block of many microwave devices. Figure 2.4(a) shows a *pn* junction just as the p and n type

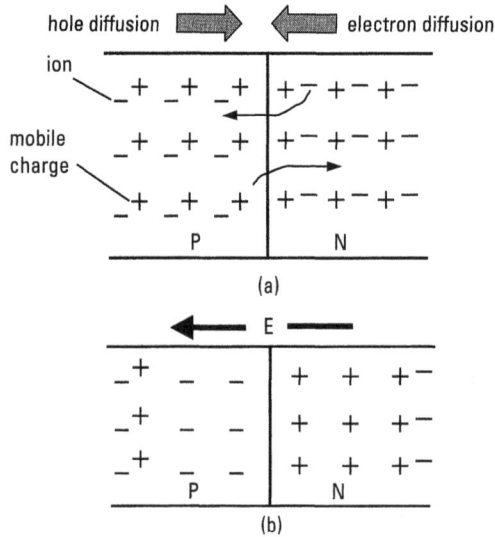

Figure 2.4 (a) *pn* junction at moment of formation has unbalanced free charge concentrations of opposite polarities. (b) *pn* junction in equilibrium: diffusion force is balanced by static field.

semiconductor materials are short-circuited together. The p-type material has a higher concentration of mobile positive charges called *holes*, and the n-type material has a higher concentration of mobile negatively charged electrons. Diffusion, a mechanical force driven by the differing particle concentrations in the p and n samples moves electrons across the junction into the p-type material, and holes into the n-type material. The resulting net positive charge in the n material and net negative charge in the p material establishes an electrostatic field across the junction that is directed against the diffusion force. Once the diffusion and electrostatic forces are balanced, charge stops moving, leaving a built-in electrostatic potential [see Figure 2.4(b)].

Our discussion of the *pn* junction tells us that unbalanced forces are required for current flow; in other words, a steady current cannot be maintained in the same direction in a closed circuit by a conservative static electric field [4]. By *conservative*, we mean that there is no mechanism by which energy can be dissipated—energy is either potential or kinetic [5]. In the *pn* junction the diffusing charges have kinetic energy, which changes to potential energy in the form of the built-in electrostatic potential. Mathematically, for an electrostatic field, we have around any closed loop the following:

$$\int_C \mathbf{E} \cdot d\mathbf{L} = \int_C \mathbf{J}/\sigma \cdot d\mathbf{L} = 0 \qquad (2.14)$$

Sources of a nonconservative field, such as batteries (which convert chemical energy to electrical energy) and electric generators (which convert mechanical energy into electric energy), can maintain unidirectional current flow in a circuit.

2.2.4 DC Ground

With this background in electrostatics, let us take a closer look at DC grounding. Ground can be defined as a source or sink of field lines, and it is often, but not always, zero volts potential in a system. Consider the three cases shown in Figure 2.5. For the single particle in Figure 2.5(a), the field lines do not actually terminate on other charges, so ground is understood to occur at an infinite distance from the particle, where the potential has decreased to zero volts. Figure 2.5(b) shows a two-charge dipole, with zero potential occurring midway between the charges. In this case, the choice of ground is arbitrary, since we could choose it to be either charge; the negative charge usually is selected. One should note that the potential of the negative charge is not zero volts. If we remove the negative charge and replace it with a perfect conductor as in Figure 2.5(c), free electrons in the conductor will move so as to eliminate all tangential electrostatic fields, and they will terminate all field lines perpendicular to the

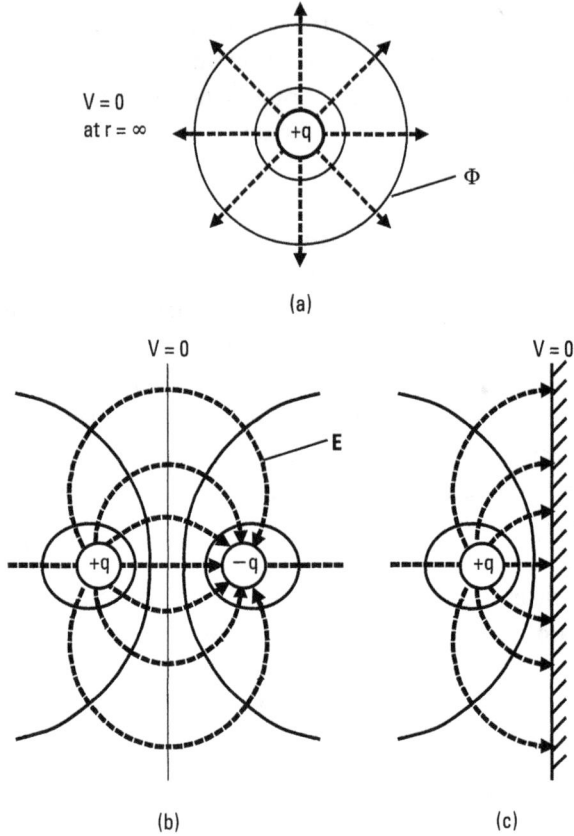

Figure 2.5 Ground: (a) at infinity for a single particle; (b) arbitrarily defined for a charge dipole; and (c) on a perfect conducting plane next to a line charge.

conductor surface. In so doing, the electrons will set up zero potential within the conductor. Thus, the perfect conductor will be ground; such a conductor is often called a *ground plane*. In summary, ground can be the potential an infinite distance from a charge, one of two or more charges in a group, or an equipotenial conductor [6].

2.3 Magnetostatics

The sources of electrostatic fields are static charges. *Magnetostatics* encompasses the theory of charges moving at a steady rate, namely steady current flow. Moving charges also generate a force at a distance, but because the charges are moving, the force is magnetic in nature. The equivalent of Coulomb's force postulate for a moving charge in a magnetic field is

$$F = q\mathbf{v} \times \mathbf{B} \qquad (2.15)$$

where q is the charge, $\mathbf{v}$ is its velocity, and $\mathbf{B}$ is the *magnetic flux density* in webers/meter^2 or Teslas.

In Figure 2.6(a) a current flows in a conducting wire. The magnetic field distribution of a steady current is *solenoidal*, wrapping itself around the wire. In other words, magnetic field lines do not originate or terminate on the charges flowing in the wire. Further, the field is perpendicular to the current flow. The right-hand rule gives the direction of the field if we know the direction of the current (thumb points in the direction of the current, curled fingers in direction of field). If we apply (2.15) to the test charge moving parallel to the current flowing in the wire of Figure 2.6(a), we see that the magnetic force moves the charge towards the wire. If we extend this situation to the two wires with parallel current flow shown in Figure 2.6(b), it is apparent that the wires will be attracted.

We can describe all magnetostatic phenomena mathematically with two equations:

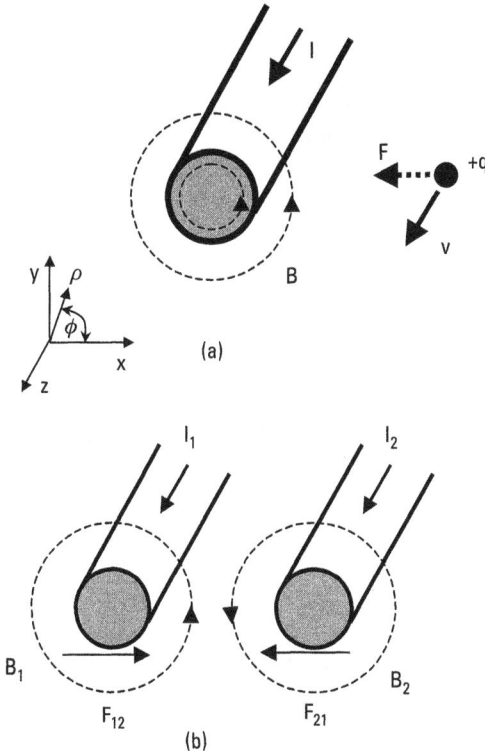

Figure 2.6 (a) A straight circular wire carrying a steady current I has a phi-directed magnetic flux. (b) Wires with aligned currents will be attracted.

$$\nabla \cdot \mathbf{B} = 0 \tag{2.16}$$

which says the lines of magnetic flux do not terminate on electrical sources, and

$$\nabla \times \mathbf{B} = \mu \mathbf{J} \tag{2.17}$$

which says that the magnetic flux lines curl around the current element $\mathbf{J}$, where $\mathbf{J}$ is the cross-sectional current density in amperes/meter^2. $\mu = \mu_r \mu_0$ is the permeability of a medium in henries/meter, with $\mu_0 = 4\pi \times 10^{-7}$ henries/meter being the permeability of free space, and μ_r is the relative permeability of the material. We can derive the continuity equation for current from (2.17):

$$\nabla \cdot \nabla \times \mathbf{B} = \mu \nabla \cdot \mathbf{J} = 0 \rightarrow \nabla \cdot \mathbf{J} = 0 \tag{2.18}$$

With the aid of Figure 2.7, we can state a key boundary condition for the magnetic field intensity $\mathbf{H}$ just outside of a perfect conductor:

$$\mathbf{n} \times \mathbf{H} = \mathbf{J} \tag{2.19}$$

where $\mathbf{n}$ is a vector normal to the surface of the conductor, $\mathbf{H} = \mathbf{B}/\mu$ in amperes/meter. In words, the surface current flowing on a conductor is normal to the induced magnetic field. Numerical electromagnetic analysis software often calculates the magnetic field in the vicinity of a conductor and uses (2.19) to calculate the current flowing on the conductor.

Inductance is an important characteristic of conductors that provide grounding. It is associated with current flowing through a wire such as that

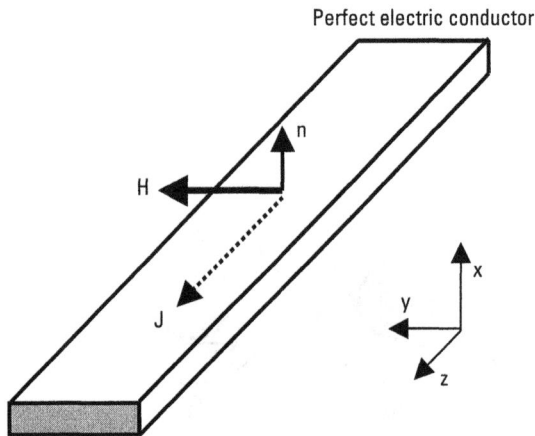

Figure 2.7 The surface current flowing on a perfect electric conductor is normal to the induced magnetic field.

shown in Figure 2.6(a). The total magnetic flux that passes through the conductor is given by the integral of the field over the cross-sectional area S of the wire:

$$\Phi = \int_S \mathbf{B} \cdot d\mathbf{s}$$

We know from (2.19) that $\mathbf{B}$ is proportional to $\mathbf{J}$, so the flux Φ must be proportional to the magnitude J of the current. The proportionality factor is called the *self-inductance* of the wire, or

$$\Phi = LJ \rightarrow L = \Phi/J \tag{2.20}$$

where the units of inductance are henries. Given the inductance of a wire, its AC reactance is $X = \omega L$, where $\omega = 2\pi \times$ frequency.

2.4 Electromagnetics

Static electricity and magnetism describe the electric and magnetic forces at a distance due to stationary charges and steady currents. Static field theory is sufficient to explain the behavior of low frequency lumped circuits and higher frequency, but electrically small, structures such as semiconductor devices. In this book, our primary interest is steady state, time-varying electromagnetics. Once the electric and magnetic fields become time-varying, they become coupled: the time-varying electric field produces a time-varying magnetic field, which produces a time-varying electric field, and so forth. This coupling is the principal behavior described by electromagnetic theory, and it makes time-varying fields uniquely different than static ones. While electrostatic fields only can store energy, electromagnetic fields can transfer energy over a distance through the phenomenon of radiation.

2.4.1 Maxwell's Equations

Coulomb's postulate, which describes the forces at a distance between charges, is also the central postulate of electromagnetic theory. Maxwell's equations follow from Coulomb's postulate and the Lorentz transformation [7]:

$$\nabla \times \mathcal{E}(x, y, z, t) = -\partial \mathcal{B}(x, y, z, t)/\partial t \tag{2.21}$$

$$\nabla \times \mathcal{H}(x, y, z, t) = \mathcal{J}(x, y, z, t) + \partial \mathcal{D}(x, y, z, t)/\partial t \tag{2.22}$$

$$\nabla \cdot \mathcal{B}(x, y, z, t) = 0 \qquad (2.23)$$

$$\nabla \cdot \mathcal{D}(x, y, z, t) = \rho(x, y, z, t) \qquad (2.24)$$

where:

- $\mathcal{E}$ is the electric field intensity in volts/meter, a three-dimensional vector quantity.
- $\mathcal{H}$ is the magnetic field intensity in amperes/meter, a three-dimensional vector quantity.
- $\mathcal{B}$ is the magnetic flux density in webers/meter^2, a three-dimensional vector quantity.
- $\mathcal{D}$ is the electric flux density in coulombs/meter^2, a three-dimensional vector quantity.
- $\mathcal{J}$ is the electric current density in amperes/meter^2, a three-dimensional vector quantity.
- ρ is the electric volume charge density in coulombs/meter^3, a scalar quantity.

Equations (2.21) and (2.22) state that a time-varying electric or magnetic field will induce a complementary time-varying magnetic or electric field. Equations (2.23) and (2.24) are drawn from our earlier discussion of static fields.

From (2.22), we can derive the general continuity equation following the approach we used to derive (2.18). If we take the divergence of (2.22) and substitute for $\nabla \cdot \mathcal{D}$ using (2.24), we get the general form of the current continuity equation:

$$\nabla \cdot \mathcal{J}(x, y, z, t) = -\partial \rho(x, y, z, t) / \partial t \qquad (2.25)$$

This equation is the basis for storage of charge at the ends of dipole antennas and at the edges of slots in ground planes.

The behavior of an electromagnetic field depends on the material in which it occurs. Free space is charge-free vacuum or air (to a close approximation in most situations). Other materials contain charged particles that interact with and modify the electromagnetic field. The *constitutive relationships* are equations that relate the field vectors within a material:

$$\mathcal{D}(x, y, z, t) = \varepsilon \mathcal{E}(x, y, z, t) \qquad (2.26)$$

$$\mathcal{B}(x, y, z, t) = \mu \mathcal{H}(x, y, z, t) \qquad (2.27)$$

$$\mathcal{J}(x, y, z, t) = \sigma \mathcal{E}(x, y, z, t) \qquad (2.28)$$

where $\varepsilon = \varepsilon_r \varepsilon_0$ is the permittivity or dielectric constant in farads/meter with ε_0 being the permittivity of free space (see Section 2.2), and ε_r is the relative permittivity of the material; $\mu = \mu_r \mu_0$ is the permeability of a medium in henries/meter, defined already in Section 2.3. σ is the conductivity of a material in siemens/meter, as defined in Section 2.2.2. In this book, the dielectric constant, permeability, and conductivity are linear, homogenous (independent of position), isotropic (invariant with direction of applied field), and not functions of the applied field.

We assume that all circuits are driven in a time-harmonic and steady state manner so that the time dependence is sinusoidal. Further, we assume that the time and space dependencies of the fields are separable into functions of time and space. Then we can write the electric field as $\mathcal{E}(x, y, z, t) = \mathrm{Re}\left[E(x, y, z)e^{j\omega t} \right]$ and similarly for the other fields, the current and the charge. If we substitute these expressions into (2.21) to (2.24), we get the time-harmonic versions of Maxwell's equations:

$$\nabla \times E(x, y, z) = -j\omega B(x, y, z) \qquad (2.29)$$

$$\nabla \times H(x, y, z) = J(x, y, z) + j\omega D(x, y, z) \qquad (2.30)$$

$$\nabla \cdot B(x, y, z) = 0 \qquad (2.31)$$

$$\nabla \cdot D(x, y, z) = \rho(x, y, z) \qquad (2.32)$$

Electromagnetic energy propagates down transmission lines and radiates from antennas into space with wave-like properties. If we take the curl of (2.29) and substitute for $\nabla \times H$ using (2.30), we get the *wave equation*:

$$\nabla \times \nabla \times E = -j\omega\mu J + \omega^2 \mu\varepsilon E \qquad (2.33)$$

where we have used (2.26) and (2.27) and suppressed the spatial dependence. Equation (2.33) contains several parameters commonly used to describe the characteristics of electromagnetic waves:

- $c_0 = 1/(\mu_0 \varepsilon_0)^{1/2}$, the speed of light in free space;

- $\lambda_0 = c_0/f$, the free-space wavelength, the distance over which the wave is periodic;
- $c = 1/(\mu\varepsilon)^{\frac{1}{2}}$, the speed of light in a medium having constitutive parameters μ and ε;
- $\lambda = c/f$, the wavelength in the same medium;
- $k = \omega(\mu\varepsilon)^{\frac{1}{2}} = \omega/c = 2\pi f/c = 2\pi/\lambda$, the wavenumber or propagation constant, which says the phase of a wave front changes by 2π radians in one wavelength of travel.

In a source-free region such as an empty waveguide or coaxial transmission line, $\mathbf{J} = 0$, and with $\nabla \times \nabla \times \mathbf{E} = \nabla(\nabla \cdot \mathbf{E}) - \nabla^2\mathbf{E} = -\nabla^2\mathbf{E}$, we can reduce (2.33) to the *Helmholtz equation*:

$$\nabla^2\mathbf{E} + k^2\mathbf{E} = 0 \qquad\qquad (2.34)$$

We solve (2.34) for the fields in source-free regions such as transmission lines and other guided wave structures.

Electromagnetic waves transfer energy, and the *Poynting vector* gives the instantaneous rate of energy flow per unit area at a point in space as [8]

$$\mathbf{P} = \mathbf{E} \times \mathbf{H} \text{ in watts/meter}^2 \qquad\qquad (2.35)$$

Equation (2.35) says that electromagnetic energy is transported by the electromagnetic field in a direction perpendicular to the plane of the electromagnetic field. This concept is fundamental to the field-based paradigm. However, because fields are *nonphysical,* mathematical descriptions of the forces exerted by charges and currents on each other, nothing physical is actually flowing through space [9]. Rather, the energy of (2.35) is being transferred over a distance from one set of sources to another with no interaction occurring in the intervening space. For example, the electrons flowing through a light bulb are the source of a radiated field that interacts with the human eye or any other photosensitive detector to generate a flow of electrons that can be used for signal processing and image recognition [10]. In a microwave oven, the same interaction excites the electrons within food, heating the food.[2]

---

2. We can say that electromagnetic field theory describes the precise relationship and behavior of the forces that occur over a distance between sources, and we can use (2.35) to calculate the energy that is transferred. But the electromagnetic interaction between two sources separated by empty space occurs over a distance, with no intervening interaction. Now, one might suggest that photons are flowing through the space between the sources. But a photon

Just as a static charge is the source of an electrostatic field, an electromagnetic source is a time-varying current or charge, which generates an electromagnetic field. Figure 2.8 shows how electrostatic and electromagnetic sources are related. If we have a line source with charge +q as in Figure 2.8(a), it creates a radially outward-directed electric field. If we now have some means to vary the

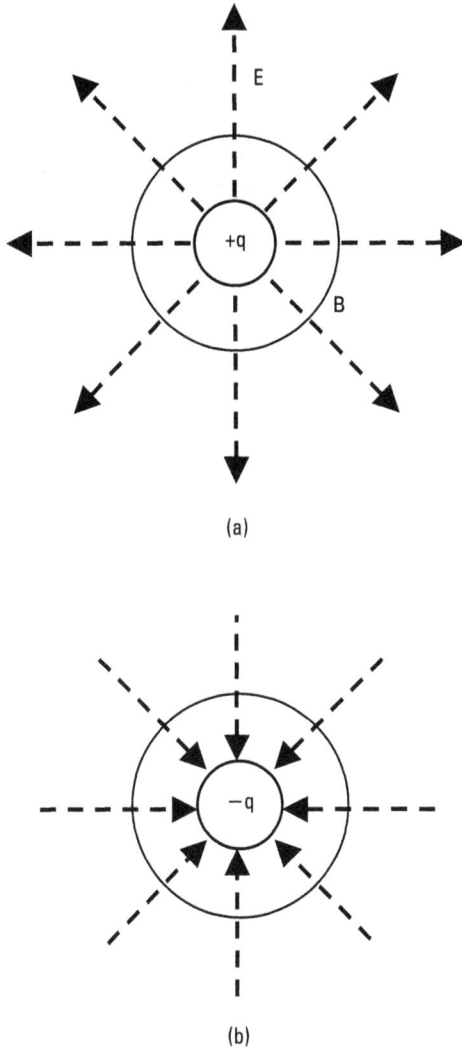

(a)

(b)

Figure 2.8 Line charge with oscillating phase causes electric field to oscillate in direction, inducing an oscillating magnetic field: (a) phase = 0°; and (b) phase = 180°.

is not a physical particle. It is a quantum or unit of electromagnetic energy, the smallest amount or bundle of energy that can be transferred at a distance.

charge polarity from $+q$ to $-q$ at a particular frequency, we have an oscillating RF source. Over time, the electric field direction will cycle from inward to outward at the frequency of the oscillating charge as shown in Figure 2.8. If the electric field is time-varying, then it must induce a time-varying magnetic field as shown in Figure 2.8, which also alternates in direction.

2.4.2   RF Ground

We reexamined the definition of DC ground as part of our discussion on electrostatics. Now, we briefly will revisit RF grounding with the aid of Figure 2.8. This oscillating charge is hardly any different than the static one in Figure 2.5(a). Recall that ground can be a source or a sink of current or charge. Thus, in Figure 2.8, even with the charge polarity continuously changing, ground is at infinity, where the potential is zero volts. Similarly, ground is the reference potential in an electromagnetic circuit, which may include conductors where field lines terminate or originate. Note that for electromagnetic structures such as transmission lines that sustain operation to 0 Hz, the electrostatic definition of ground must apply also. These structures include transverse electromagnetic (TEM) and quasi-TEM transmission lines (see Chapter 3), antennas with TEM and quasi-TEM radiation modes (see Chapter 6), and microwave transistors (see Chapter 5).

2.5   Electromagnetic Radiation and Antennas

As preparation for Chapter 6, on antennas and grounding, we will discuss one last topic—radiation. According to the Institute of Electrical and Electronic Engineers (IEEE), *radiation* is "the emission of energy in the form of electromagnetic waves" [11]. Fundamentally, the physical mechanism that causes radiation is *acceleration* of charged particles. Since particles do not accelerate in electrostatics or magnetostatics, radiation cannot occur at 0 Hz. One example of acceleration is the simple, straight-line increase in velocity that occurs in an atom smasher. For a nonrelativistically accelerating electron, Sommerfeld shows that energy is lost to radiation at the rate $e^2 a / 6\pi\varepsilon_0 c^3$, where a is acceleration, c is the speed of light, and e is the electron charge [12]. Of more interest to us is a stationary, but time-varying charge or current distribution, which also radiates. In addition, radiation occurs at circuit discontinuities, which force currents to change direction (a form of acceleration). For instance, time-varying current flowing on a transmission line encounters a discontinuity such as an open end that causes an abrupt change in its direction and velocity. The discontinuity causes the current to decelerate and radiate. A radiating structure within an electromagnetic circuit acts as a load and decreases the potential of current that flows across its terminals.

Reciprocity requires that if a charged particle can transmit or lose energy to radiation, then it can receive or gain energy by intercepting radiation from another particle. For example, a source current on an antenna can radiate and induce currents to flow on a receiving antenna. In this way, energy is transferred through radiation [13].

Poynting's vector, (2.35), tells us that energy is transmitted by an electromagnetic field. Both an electric and a magnetic field must be present, and they must be coupled. If either the electric field or the magnetic field is zero, no energy can be transmitted. In free-space, far from the source of radiation, the energy in the electromagnetic wave spreads over a spherical wave front, and thus the power in the wave at any one point on the wave front falls off as $1/r^2$, where r is the distance from the source. The electric and magnetic fields have magnitudes proportional to $1/r$.

Earlier, we explained that although energy can be viewed as transmitted via a radiating electromagnetic field, charges and currents are the source of the field. As a consequence, radiating structures such as antennas are constructed from conductors on which currents flow. Dielectrics such as lenses may be part of the antenna structure, but they only serve to shape the radiating electromagnetic field. Antenna designers optimize a radiating structure's geometry to maximize its ability to transfer to free-space the energy at its input terminals. Figure 2.9 shows a dipole antenna with its radiated fields and currents. The primary radiating mechanism for the dipole is the abrupt termination of the current flow at the ends of the conductors making up its arms. The continuity equation (2.25) requires that charges of opposite polarity be stored at the ends. If the source is oscillating, then the charges will switch polarity at the same frequency, and the dipole will radiate an electromagnetic field. The length of the dipole determines the separation of the charges, and thus how the radiation from the two ends combines in space. When we choose the dipole to be about half of one wavelength, the dipole's input impedance is resonant (pure real-valued), and the energy radiated is maximized.

Because the edge of a sheet conductor interrupts the flow of current also, charge accumulates there, as shown in Figure 2.10(a). If another conducting sheet edge is placed in close proximity, charge of opposite polarity accumulates on that sheet's edge. The two charged edges establish an electric field across the gap or slot between the two sheets. If the current source is oscillating, then the slot can radiate an electromagnetic field. In general, anytime an aperture interrupts the flow of time-varying current on a conductor, the accumulated charge on the edge of the aperture will excite an electromagnetic field within the aperture. As with the dipole antenna, the aperture's dimensions will determine its efficiency as a radiator. The slot and dipole are dual antennas, meaning that the

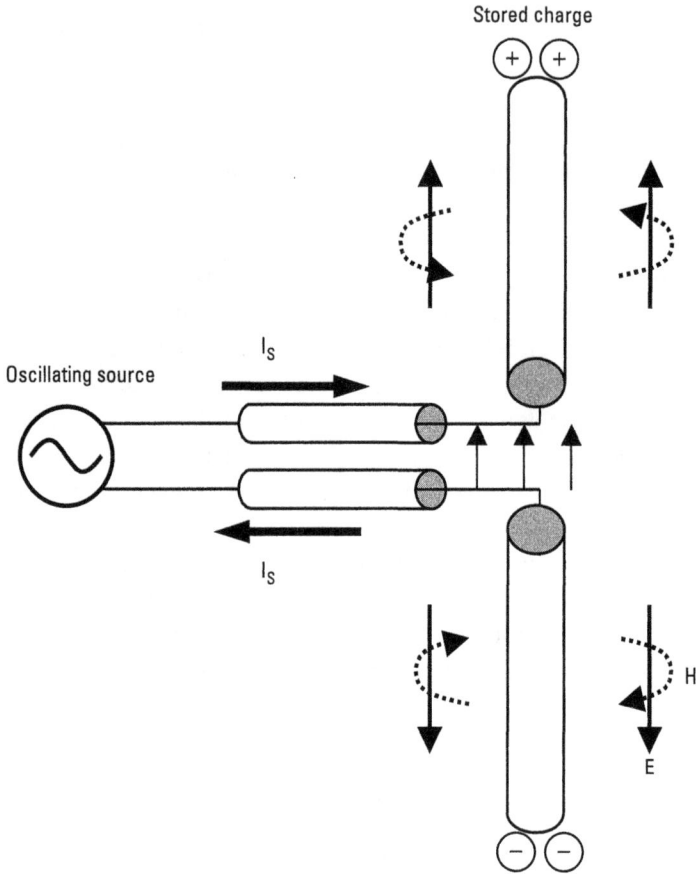

Figure 2.9 Dipole antenna and its source.

slot's electric field distribution has the same shape as the dipole's magnetic field, and vice versa. Consequently, for a thin slot in a ground plane, resonance and maximum radiation efficiency occur when the slot length is about one-half wavelength.

A radiating source such as an antenna interacts with other sources with an intensity that varies both spatially and with frequency. *Intensity* is the power density in the radiated field, and it is proportional to the square of the amplitude of the field. As we move radially outward from the antenna, the radiated field assumes a fixed variation with angle in a relative sense that is independent of distance from the source. This region extends to infinite distance, and it is called the *far field*. In the far field, an electromagnetic wave usually approximates the form of a *uniform plane wave*, for which the electric and magnetic field components lie in the same plane and are perpendicular to the direction of wave propagation. In contrast, the region near the source is characterized by energy storage,

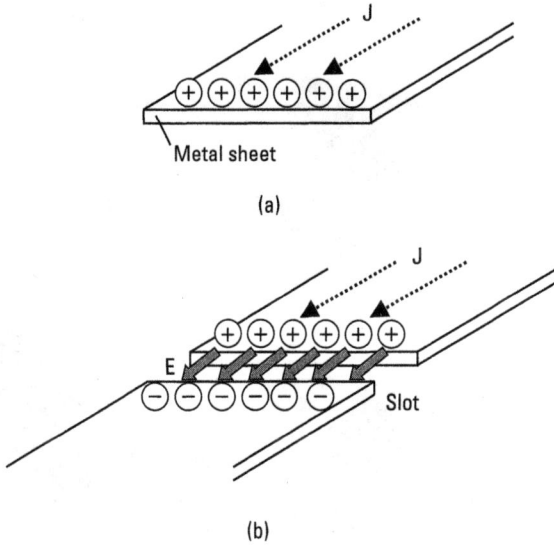

Figure 2.10 Current **J** flowing on a sheet conductor: (a) charge builds up at edge; and (b) when a second sheet is brought nearby, an electric field forms in the slot.

and it is termed the *near-field* region. The transition between near and far-field regions is gradual; however, a distance from an antenna of $2D^2/\lambda$, where D is the largest dimension of that antenna, generally is considered to be in the far field.

We usually compare the radiated field of most antennas and radiating structures with that of an ideal *isotropic* source, which has a constant intensity in all directions. A dipole antenna like that in Figure 2.9 is an *omnidirectional* antenna in that the intensity of its radiated field is constant around its axis.

A *radiation* or *power pattern* is a spatial description of a radiating structure's intensity at a single frequency. Figure 2.11 shows a three-dimensional view of a dipole's far-field radiation pattern throughout all space. We often study two-dimensional pattern cuts along key planes intersecting an antenna's three-dimensional pattern. Figure 2.12 shows a *polar plot* of the far-field radiation intensity versus angle of a dipole antenna in the plane of its electric field, with power in decibels normalized so that the peak of the pattern is 0 dB. The power level at any single angle (theta, phi) normalized to that of an isotropic source is called the *directivity* at that angle. The directivity of a perfectly matched isotropic radiator is 0 dBi at all angles. All real antennas have directivity that exceeds 0 dBi at one or more angles, and negative directivity at others. An antenna's *gain* is equal to its directivity (in dBi) less losses (in decibels) due to input mismatch and resistance in the antenna. A two-dimensional radiation pattern also can be plotted as a Cartesian plot with power, directivity or gain on the *y*-axis and observation angle on the *x*-axis.

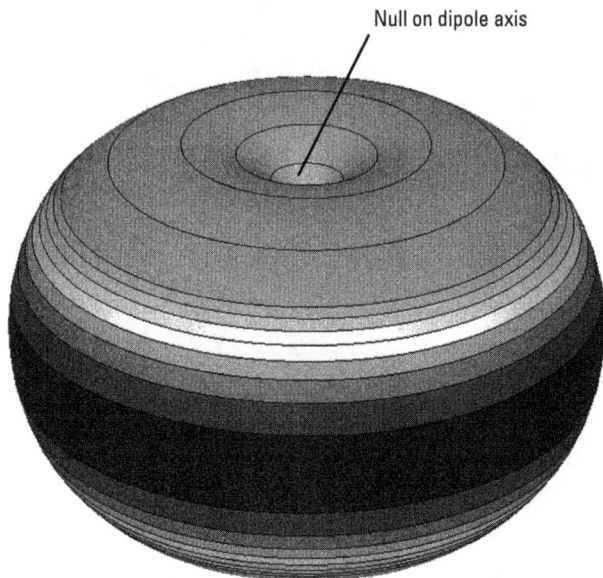

Figure 2.11 Three-dimensional, omnidirectional, far-field radiation pattern of a dipole antenna.

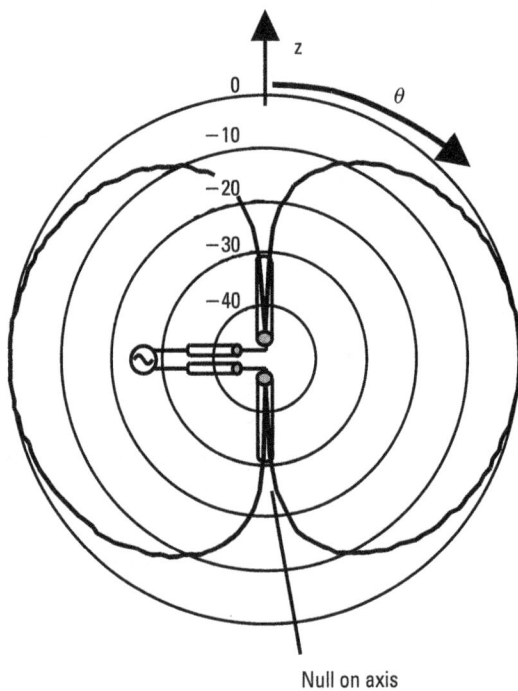

Figure 2.12 Polar radiation pattern of a dipole antenna.

References

[1] Elliott, R. S., *Electromagnetics*, New York: IEEE Press, 1993, p. 116.

[2] Jordan, E. C., *Electromagnetic Waves and Radiating Systems*, Upper Saddle River, NJ: Prentice Hall, 1950, p. 61.

[3] Balanis, C. A., *Advanced Engineering Electromagnetics*, New York: John Wiley & Sons, 1989, p. 61.

[4] Cheng, D. K., *Fields and Wave Electromagnetics*, Reading, MA: Addison-Wesley, 1983, pp. 177–178.

[5] Jordan, E. C., *Electromagnetic Waves and Radiating Systems*, New York: Prentice Hall, 1950, p. 34.

[6] Harrington, R. F., *Introduction to Electromagnetic Engineering*, New York: McGraw-Hill, 1958, p. 198.

[7] Elliott, R. S., *Electromagnetics*, New York: IEEE Press, 1993, pp. 264–270.

[8] Elliott, R. S., *Electromagnetics*, New York: IEEE Press, 1993, pp. 285–291.

[9] Panofsky, W. K. H., and M. Phillips, *Classical Electricity and Magnetism*, Second Edition, Reading, MA: Addison-Wesley, 1962, p.1.

[10] Kraus, J. D., *Electromagnetics*, 3rd ed., New York: McGraw-Hill, 1973, p. 716.

[11] *Standard Dictionary of Electrical and Electronics Terms*, 4th ed., IEEE Std 100-1988, New York: IEEE, 1988, p. 773.

[12] Sommerfeld, A., *Electrodynamics*, New York: Academic Press, 1952, p. 293.

[13] Stutzman, W. L., and G. A. Thiele, *Antenna Theory and Design*, New York: John Wiley & Sons, 1981, Section 1.2.

3

Transmission Lines, Waveguides, and Passive Circuits

At microwave frequencies, a conducting wire is a poor purveyor of electrical energy. Instead we use a transmission line or waveguide to transfer electromagnetic waves between a generator and load. In contrast to wires, the conductors of a transmission line are configured in a specific geometrical relationship to make this energy transfer as efficient as possible. In this chapter, we review transmission line and waveguide theory and examine the characteristics of multiwire, planar, coaxial, and waveguide guided structures. We then describe how discontinuities and impedance in the ground path of transmission lines hinder the flow of current and cause loss of power and unwanted radiation. We also compare DC and RF short circuits for terminating transmission lines and other passive circuits. The chapter concludes with an extensive discussion of techniques for grounding multilayer, mixed signal RF printed circuit boards and passive surface mount components.

3.1 Fundamental Theory

There are two ways to transfer electromagnetic energy between an electrical source and a load that are separated in space. As shown in Figure 3.1, we can use either antennas or a transmission line. In the first method, we use the source to drive an antenna, which focuses the peak intensity of the radiated electromagnetic field in the direction of the load. We receive the signal on another antenna as currents induced to flow by the transmitting antenna's electromagnetic field. These currents flow to the load. For example, a high power transmitter drives a

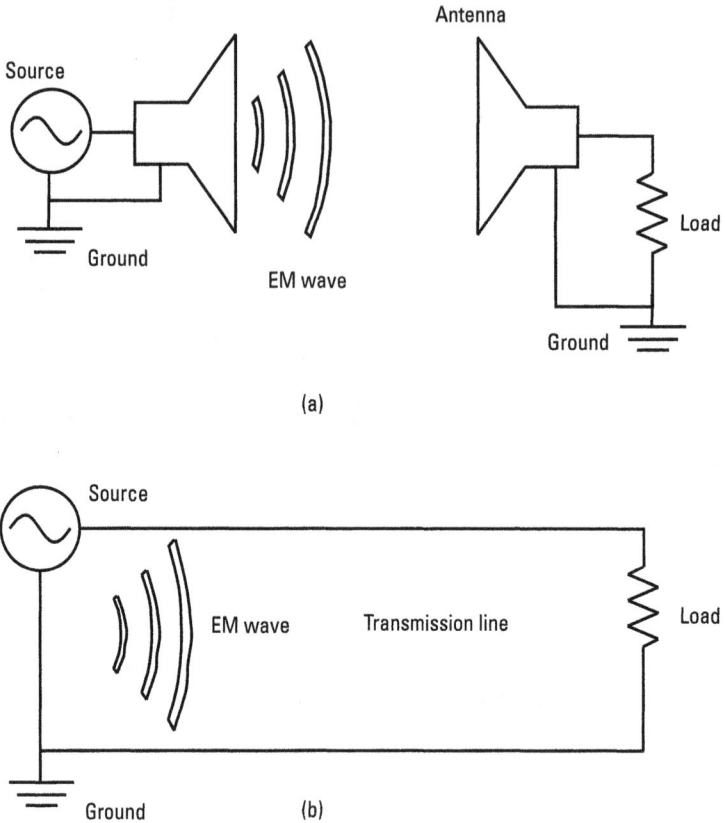

Figure 3.1 A point-to-point communications link uses (a) highly directive antennas or (b) transmission line to transfer electromagnetic energy over long distances.

radio station's vertical monopole antenna, which broadcasts radio waves over the Earth's surface. A person desiring to receive the signal can place an antenna in the path of the electromagnetic wave. Currents induced on the receiving antenna will flow into his receiver and be down converted to signals in the audio band.

For many applications, an antenna is a very efficient and inexpensive device for transmitting or receiving microwave energy across space. Since no current conducts through free-space, we need concern ourselves only with grounding at the transmitter and receiver, not in the space between. The antennas of a point-to-point link like that shown in Figure 3.1(a) direct signals between two specific geographic points. Such antennas are often electrically large to tightly confine the electromagnetic energy.

As an alternative, we can replace the antennas with a transmission line [see Figure 3.1(b)], a structure made from conductors and/or dielectrics that constrains and guides electromagnetic waves over a distance. There are two types of transmission lines: (1) those that confine the electromagnetic field using conductors on which flow the currents that are the sources of the field, and (2) those that are conductor-less, behaving in a sense like antennas, in that they shape and confine the electromagnetic waves radiated from a source but do not conduct its current to the load. Figure 3.2 shows an example of the first type, a *coaxial transmission line*, which is constructed from two concentric circular conductors separated by an insulator. At high frequencies, the signal current flows on or very near the surface of the inner conductor to the load, while the ground current returns to ground along the inner surface of the outer conductor. Figure 3.3 shows a commonly used, conductor-less transmission line, the optical fiber. This particular fiber consists of two concentric dielectrics called the core and cladding. The source currents flowing on a light source such as a light-emitting diode send a light-wave into one end of the fiber, which induces currents to flow on a detector (the load) at the other end. The dielectric constants of the core and cladding are greater than that of air. Consequently, if the propagating electromagnetic wave enters the fiber at an angle that is nearly parallel with the fiber axis, Snell's law requires that the light-wave be constrained inside the fiber by total internal reflection [1]. Such an optical link resembles the point-to-point free-space link in Figure 3.1(a) more than the transmission line in Figure 3.1(b),

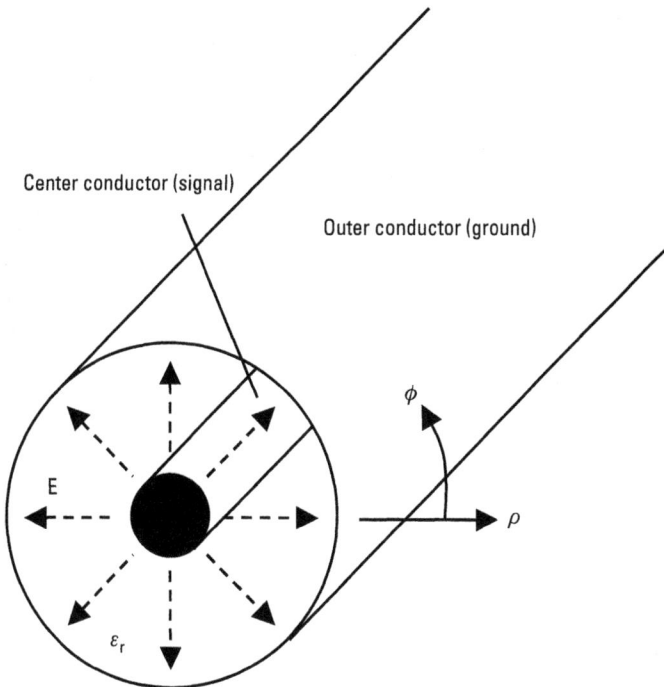

Figure 3.2 The conductors of a coaxial transmission constrain the electric field. ε_r is the dielectric constant of the insulator.

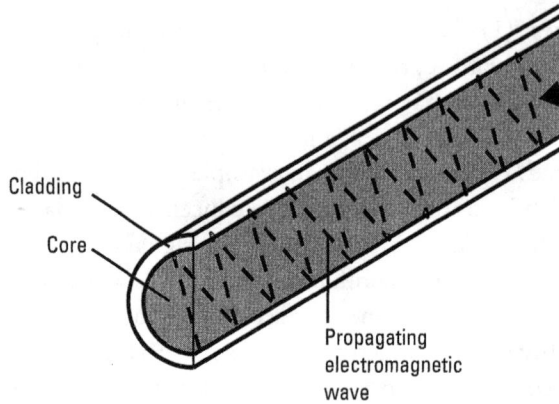

Figure 3.3 An optical fiber, a conductor-less transmission line made from two or more dielectric layers, uses total internal reflection to constrain the electromagnetic field and transfer energy.

since no current flows in the energy transmitting medium, the optical fiber. As with the free-space link, there is no signal or ground path between the source and load, and grounding problems cannot occur between them. In this chapter we will focus on transmission lines and waveguides that use conductors.

We can describe a standard transmission line having a uniform cross-section with the aid of Cartesian or cylindrical coordinates and solve the Helmholtz equation (2.34) for the field distributions that can arise. The mathematical details of the solution are straightforward and can be found in the references. The solution tells us that for every transmission line or waveguide, an infinite number of unique cross-sectional field configurations or modes can exist, with each mode having a propagational dependence given by $e^{-\gamma mnZ + j\omega t}$. γ_{mn} is the *propagation constant* of the mnth mode, where m and n are indices, typically zero or greater, that indicate the order of field variation in each of the two-dimensional cross-sectional coordinates. A value of zero for a particular index means the field distribution is constant in the associated dimension. Above a certain frequency called the *cutoff frequency* the propagation constant is purely imaginary, $\gamma_{mn} = j\beta_{mn}$, and the mode propagates down the length of the transmission line or waveguide as an electromagnetic wave with very low loss. Below the cutoff frequency, the propagation constant is purely real, $\gamma_{mn} = \alpha_{mn}$, and the mode is highly attenuative or evanescent, decaying to zero amplitude in a very short distance. The mode with the lowest cutoff frequency is termed the *dominant mode*. Generally, transmission lines are operated at frequencies such that only the dominant mode can propagate, all other modes being evanescent.

An important parameter for all transmission lines and waveguides is the mode *wave impedance,* the ratio of cross-sectional fields in the line

$$Z_w = E_u/H_v = -E_v/H_u \tag{3.1}$$

where u and v are the transverse or cross-sectional coordinates, such as (x, y) and (ρ, ϕ). Impedance matters most when we are analyzing propagation problems. For example, suppose we connect a source and a load each to its own transmission line, with the transmission lines having different impedances, and then we form a junction by connecting the two transmission lines. The impedance mismatch will cause a portion of the energy in a wave incident at the junction to be reflected and returned to the source. This energy not transmitted to the load on the other side of the junction is called *mismatch loss*.

In Figure 3.4, an electromagnetic source sends a voltage wave down a transmission line that is terminated in a load with complex impedance Z_L. In general, the load impedance is not matched to the transmission line, so the load absorbs some energy from the incident wave and reflects some back towards the source. Because the incident and reflected waves propagate on the transmission line simultaneously, and their sum varies as a function of the longitudinal coordinate z, we write the voltage phasor between the conductors and the current phasor flowing on the signal conductor as

$$V(z) = V^+ e^{-j\beta z} + V^- e^{+j\beta z} \tag{3.2}$$

$$I(z) = I^+ e^{-j\beta z} + I^- e^{+j\beta z} \tag{3.3}$$

where V^+, V^-, I^+, I^- are complex mode voltage and current amplitude coefficients related to the *characteristic impedance*, which is defined as: $Z_0 = V^+/I^+ = -V^-/I^-$. The load is located at $z = 0$, and we usually wish to

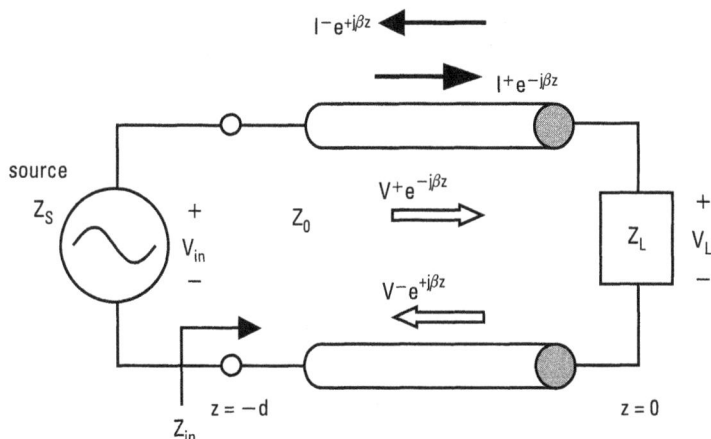

Figure 3.4 A source driving a transmission line terminated in a load.

know the *input impedance* $Z_{in}(z) = V(z)/I(z)$ at any z coordinate between the source and the load. Given the boundary condition $Z_L = V(0)/I(0)$, it is a simple matter to use (3.2) and (3.3) to derive an equation for the input impedance of a loaded transmission line a distance z from the load:

$$Z_{in} = Z_0 \frac{Z_L + jZ_0 \tan \beta z}{Z_0 + jZ_L \tan \beta z} \tag{3.4}$$

where $\beta = 2\pi/\lambda$ is the propagation constant of the mode. Equation (3.4) helps explain why electrical phase length, $\beta z = 2\pi z/\lambda$ radians, is important at microwave frequencies but not at low frequencies. For instance, consider a transmission line that is 1 foot long. At low frequencies approaching 0 Hz, the wavelength is extremely long, and β, which is inversely proportional to wavelength, is almost zero. Equation (3.4) simplifies to $Z_{in} = Z_L$, so that neither the length nor the characteristic impedance of the transmission line are relevant; the input impedance is equal to the load impedance. In contrast, at 10 GHz the free-space wavelength is about 1 inch (2.5 cm), and evaluation of (3.4) reveals that the input impedance generally is complex-valued, varying with distance from the load, and not matched to the transmission line when $Z_L \neq Z_0$.

If we are given the load impedance and the characteristic impedance, the *reflection coefficient*, V^- of the reflected wave divided by V^+ of the incident wave, is

$$\Gamma = (Z_L - Z_0)/(Z_L + Z_0) \tag{3.5}$$

where Γ, Z_L, and Z_0 may be complex valued. A commonly used parameter, *return loss* (RL), the ratio of reflected power to incident power in decibels, is defined in terms of the reflection coefficient magnitude as

$$RL = -20 \log_{10}(|\Gamma|) \tag{3.6}$$

Transmission lines form the electromagnetic interface for nearly all microwave circuits. Those circuits with one input and one output port are called *two-ports*. An example is a filter with coaxial transmission lines at its input and output ports. We can characterize such a circuit by applying an electromagnetic wave at one port, measuring the phase and amplitude of the reflected wave, and measuring the phase and amplitude of the wave that exits the other port. Figure 3.5 shows a two-port network with incident (a_1, a_2) and outgoing (b_1, b_2) waves at each port. The waves are related by the *scattering parameters* as [2]

$$b_1 = S_{11}a_1 + S_{12}a_2 \tag{3.7}$$

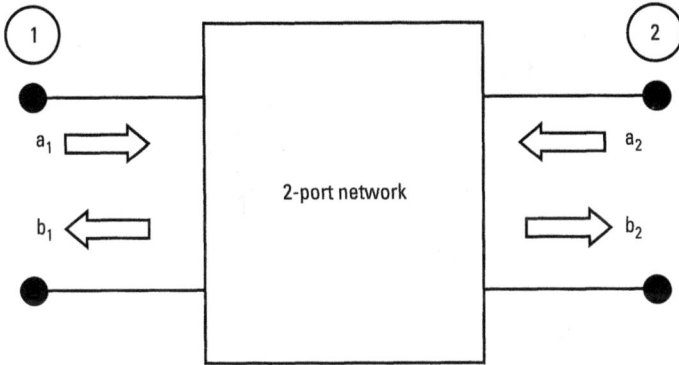

Figure 3.5 Voltage wave representation of a two-port network.

$$b_2 = S_{21} a_1 + S_{22} a_2 \qquad (3.8)$$

where S_{11} is the reflection coefficient at port 1 (b_1/a_1) with port 2 terminated in a matched load $(a_2 = 0)$; S_{22} is the reflection coefficient at port 2 (b_2/a_2) with port 1 terminated in a matched load $(a_1 = 0)$; S_{12} is the transmission coefficient from port 2 to port 1 (b_1/a_2) with port 1 terminated in a matched load $(a_1 = 0)$; S_{21} is the transmission coefficient from port 1 to port 2 (b_2/a_1) with port 2 terminated in a matched load $(a_2 = 0)$.

3.2   Coaxial Transmission Line

Coaxial transmission line is popular for its wide bandwidth, high resistance to electromagnetic interference (EMI) and low cost. Figure 3.2 shows a coaxial transmission line with its dominant mode electric field configuration. The magnetic field is distributed in concentric circles about the center conductor. This electromagnetic field distribution is known as the *transverse electromagnetic* (TEM) mode since both electric and magnetic fields are transverse to the direction of propagation. This mode is also termed a *differential mode*, because the signal current flowing on the inner conductor is directed opposite to the ground current flowing on the outer conductor. The TEM mode has several unique characteristics: (1) at least two unconnected conductors and a single insulating material are required for it to exist; (2) its cutoff frequency is 0 Hz; (3) it has only two field components (E_ρ and H_ϕ for the coaxial line) aligned with the transverse coordinates, no longitudinal (z-directed) electric or magnetic field component (E_z or H_z); and (4) its propagation constant is $k = k_0 \varepsilon_r^{1/2}$, the wavenumber in vacuum multiplied with the square root of the relative dielectric constant of the insulator. Because the TEM mode cuts off at 0 Hz, we can solve for the electric and magnetic fields using static theory, and we can define the outer conductor as ground. We can define the *characteristic impedance* in terms of the static voltage and current on the transmission line:

$$Z_c = \frac{V(z)}{I(z)} = \frac{V_0 e^{-jkz}}{I_0 e^{-jkz}} = \frac{-\int_{S_1}^{S_2} \mathbf{E} \cdot d\mathbf{l}}{\int_{S_2} \mathbf{H} \cdot d\mathbf{l}} \qquad (3.9)$$

The characteristic impedance of the coaxial transmission line is given by

$$Z_{coax} = (\eta/2\pi) \ln(OD/ID) \qquad (3.10)$$

where *ID* and *OD* are the diameters of the inner conductor and inside surface of the outer conductor. η is the *intrinsic impedance* of the coaxial insulator, given by $(\mu_r/\varepsilon_r)^{\frac{1}{2}} \eta_0$, and η_0, the intrinsic impedance of a vacuum, is equal to 120π or 377 ohms. For commonly used nonmagnetic insulators, such as air and Teflon, $\mu_r = 1$. The ratio of the conductor diameters often is chosen so that the characteristic impedance is 50 ohms. From (3.10), we see that the characteristic impedance of the TEM mode is independent of frequency, a feature that enables broadband interconnections between TEM transmission lines as discussed in Chapter 4.

For a particular application, coaxial line is usually sized to operate within its dominant TEM mode bandwidth, the frequency range between 0 Hz and the cutoff frequency of the first higher-order TE mode, which is given approximately by the formula [3]

$$f_{c1} \cong \frac{2c}{\pi(OD + ID)} \qquad (3.11)$$

Recall that the Poynting vector implies that power flows in the transmission line's field, not in the conductors. However, we emphasize that the currents that are responsible for fields reside on the conductors [4]. Consequently, if we interrupt the current flow on any of the conductors, there will be a corresponding discontinuity in the electromagnetic field and a change in impedance at that point, causing mismatch loss and possibly radiation. For example, Figure 3.6 shows a coaxial transmission line with an air gap in its center conductor. The gap is capacitive, since it separates two metal surfaces, the two ends of the center conductor. The capacitance is approximated well by $C_{gap} = \varepsilon_0 \pi (ID/2)^2 / g$, where g is the gap width and *ID* is the diameter of the coax center conductor. At 0 Hz, no DC power can be transmitted across the gap unless the voltage difference is sufficient for electrons to conduct (arc)

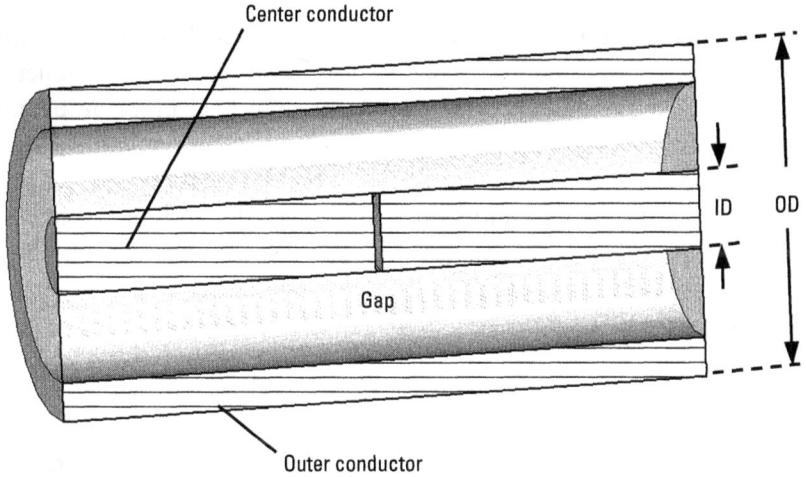

Figure 3.6 Coaxial transmission line with a capacitive gap in the center conductor.

across. Above 0 Hz, the currents on one side of the gap may radiate power to those on the other side. The series impedance of the gap is $1/j\omega C_{gap}$, so as frequency increases, the reactance decreases and the mismatch loss decreases also. Figure 3.7 plots the *insertion loss*, the ratio of power bridging the gap to that incident, calculated as $20\log_{10}|S_{21}|$ and the return loss of a 50-ohm RG-316 coaxial line [0.022-inch (0.57 mm) *ID*, 0.072-inch (1.83 mm) *OD*, $\varepsilon_r = 2.03$] with a 0.002-inch (0.05 mm) gap, which can be obtained by solving for the scattering parameters of the series reactance $1/j\omega C_{gap}$ in a transmission line [5]. Clearly, the mismatch is severe up to 50 GHz, so a break in the center conductor of a coaxial line would be easy to detect.

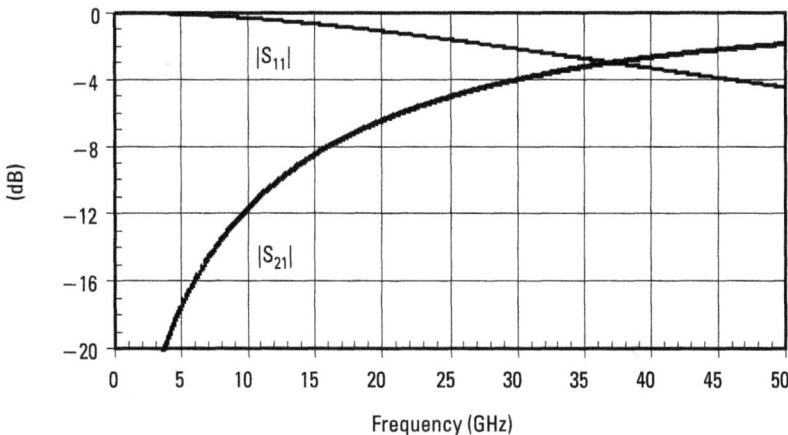

Figure 3.7 Return loss $|S_{11}|$ and insertion loss $|S_{21}|$ of a 50-ohm RG-316 coaxial transmission line with a 0.002-inch (0.051 mm) capacitive gap in the center conductor.

Of more interest to us is the effect of an interruption in the ground current path caused by a gap in the outer conductor. Unlike a break in the inner conductor, a break in the outer conductor is a potential EMI problem, because it enables interaction between the signal inside the coaxial line and the outside world. Figure 3.8 shows a break that extends around the entire outer conductor, and Figure 3.9 plots the return loss and insertion loss for an RG-316 cable with such a gap as solved with a numerical electromagnetic solver. This problem is more insidious than a break in the inner conductor, because the degradation in performance is much less pronounced. Above 9 GHz, the insertion loss is only 1

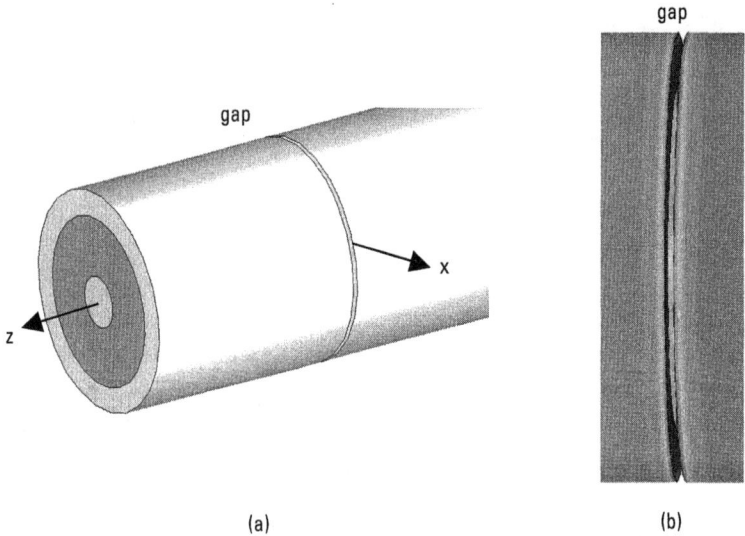

(a)   (b)

Figure 3.8 (a) Coaxial transmission line with a gap around the circumference of the outer conductor. (b) Electric field penetrates to outside the shield.

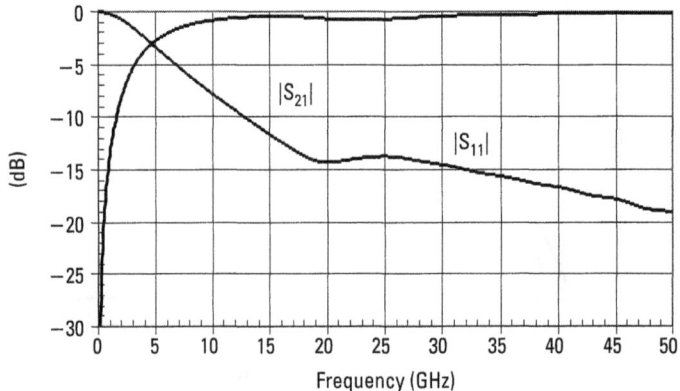

Figure 3.9 Return loss $|S_{11}|$ and insertion loss $|S_{21}|$ of a RG-316 coaxial transmission line with a 0.002-inch (0.051 mm) circumferential gap in its outer conductor.

dB; and above 32 GHz, the return loss exceeds 15 dB. If the cable were embedded between amplifiers, it would not be easy to detect a break in its ground because the slight difference in system gain would be hard to measure, and the degraded return loss of the cable would be masked by that of the amplifiers.

As the circumference of the gap approaches a wavelength, the gap's radiation efficiency increases. Since the coaxial line is an axially symmetric structure, we would expect the radiation pattern of the gap to be omnidirectional. Figure 3.10 shows an electric field pattern cut through the *x-z* plane of the transmission line. If we rotate this pattern around the *z*-axis, we get a dipole-like pattern with nulls on the axis parallel to the coaxial line. At 5 GHz—the frequency of the pattern plotted in Figure 3.10—the circumference of the outer conductor is about one-quarter wavelength. Because the radiation pattern is omnidirectional, interaction potentially can happen with any susceptible transmission line or antenna that is in the vicinity. A coaxial cable with such gaps intentionally cut periodically along its length can also be used as leaky wave antenna for communication inside tunnels and other enclosed structures.

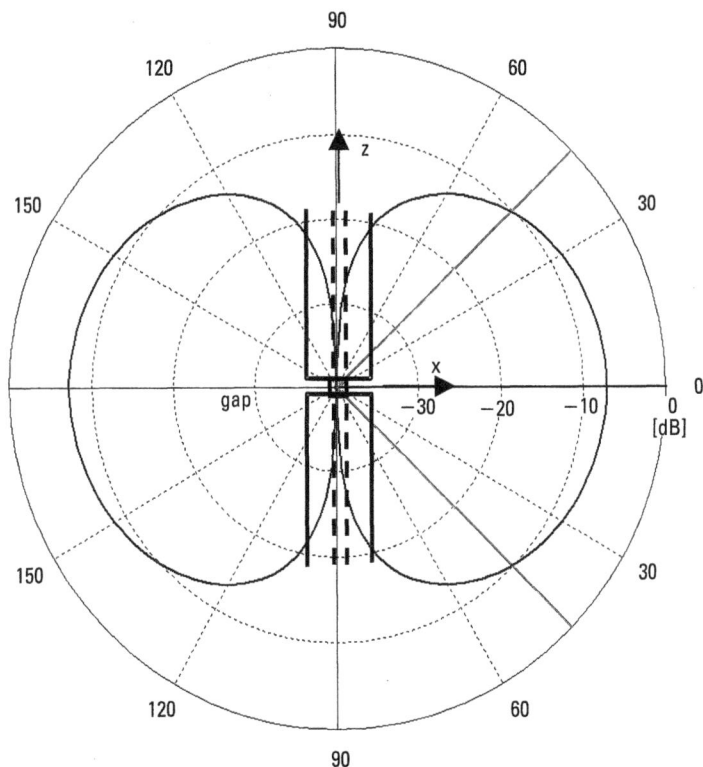

Figure 3.10 Radiation pattern of an RG-316 coaxial cable with a 0.002-inch (0.051 mm) circumferential gap, 5 GHz.

A noncircumferential break like the semicircular slot in the ground conductor shown in Figure 3.11 is even more difficult to detect. Because this slot does not extend entirely around the coaxial line, DC current will flow essentially unimpeded. Since the slot forms an aperture surrounded by a single conducting surface, the dominant mode in the aperture will have a cutoff frequency greater than 0 Hz. Consequently, the radiation efficiency will be low until the slot length reaches the cutoff wavelength, about one-half of a free-space wavelength. For the RG-316 cable in the figure, the slot's arc length is about 0.25 inch (0.63 cm), which gives a resonant frequency near 24 GHz. As Figure 3.12 shows, the return loss of the transmission line with such a discontinuity is nearly as good as

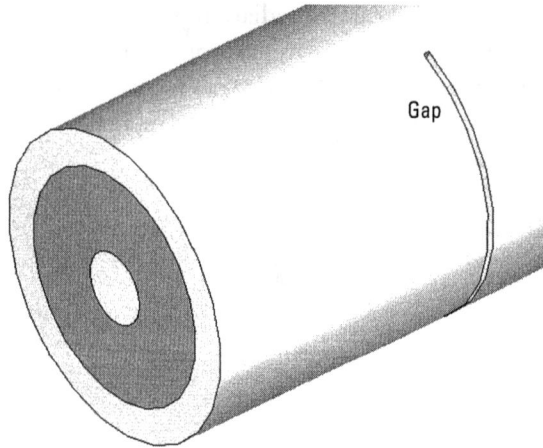

Figure 3.11 Coaxial transmission line with a semicircumferential gap in the outer conductor.

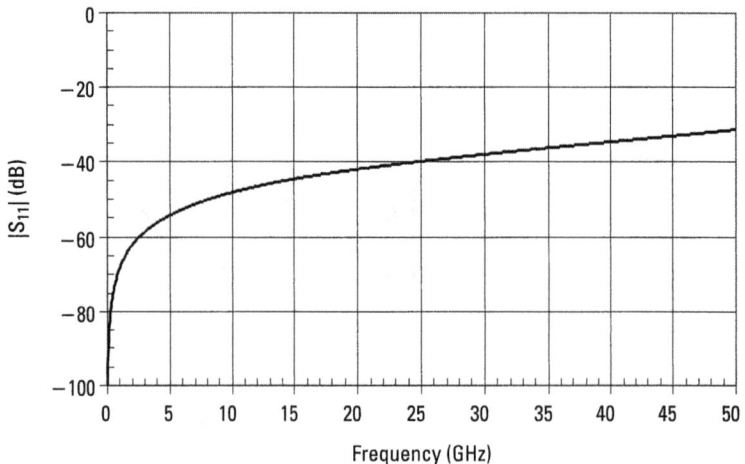

Figure 3.12 Return loss $|S_{11}|$ of a RG-316 coaxial transmission line with a 0.002-inch (0.051 mm) semicircumferential gap in its outer conductor.

that of an undamaged coaxial line. Figure 3.13 plots a radiation pattern of the coaxial line viewed end on. Although the slot extends only part way around the outer conductor, the radiated field permeates in all directions. The gain of the pattern on the side without the slot is just a few decibels below the peak gain.

Other less serious coaxial line ground path problems occur when the quality of the conductor metallization is poor. In particular, resistance in either the signal or ground conductor will dissipate power. If the outer conductor is too thin, the electromagnetic field can penetrate it. The *skin depth* of a conductor is given by

$$\delta = \left(\pi f \mu_r \mu_0 \sigma\right)^{-\frac{1}{2}} \tag{3.12}$$

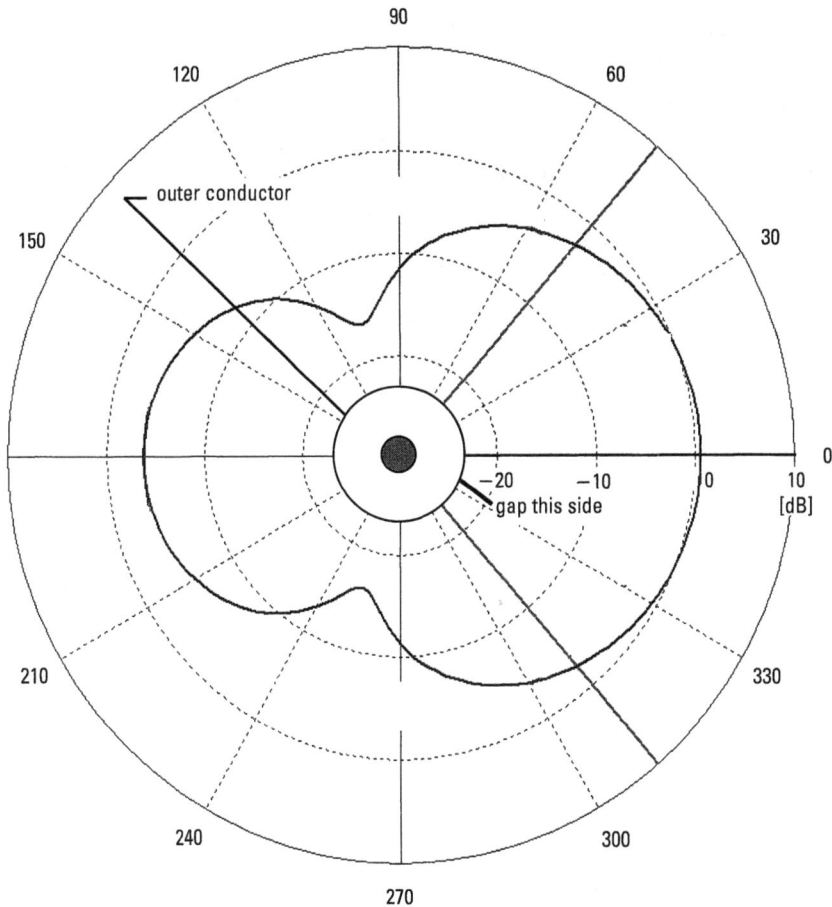

Figure 3.13 Radiation pattern of an RG-316 coaxial cable with a 0.002-inch (0.051 mm) semicircumferential gap, 20 GHz.

where f is the frequency, $\mu_r\mu_0$ is the permeability of the conductor, and σ is its conductivity. In one skin depth, the electric field is attenuated by $1/e$, which corresponds to -8.7 dB in power. Figure 3.14 plots the skin depth for several common conductors. As expected, the skin depth increases as frequency approaches 0 Hz. At frequencies above 1 GHz, the skin depth of metals is less than 0.0001 inch (2.5 microns); the outer conductor thickness of a coaxial line is much greater. However, many conductors are plated with nickel and then a layer of gold. Nickel is a lossy metal, so sufficient gold must be deposited to keep any RF current from flowing in the nickel. Three or more skin depths of gold are sufficient as $1/e^3 = 0.05$ or 95% of the energy is contained within the gold layer.

3.3 Wire Transmission Lines

The signal and ground currents of *multiconductor* transmission lines such as the coaxial line each flow on a separate conductor. The coaxial line is an *enclosed transmission line*: the outer conductor completely confines the electromagnetic field and permits no interaction with sources outside. Because of this property, coaxial transmission line finds use in interconnecting microwave modules and systems where high isolation is needed. However, the weight, size, and cost of coaxial line make it undesirable for use within individual microwave modules and at lower frequencies. Figure 3.15 shows three commonly used *open conductor*, wire transmission lines. Like coaxial line, the dominant mode of each is the TEM. Unlike the coaxial line, the electromagnetic field of an open conductor transmission line, although concentrated near the conductors, decays gradually away from the wires, finally reaching zero at an infinite distance. The signal and ground conductor diameters and their separation determine the characteristic

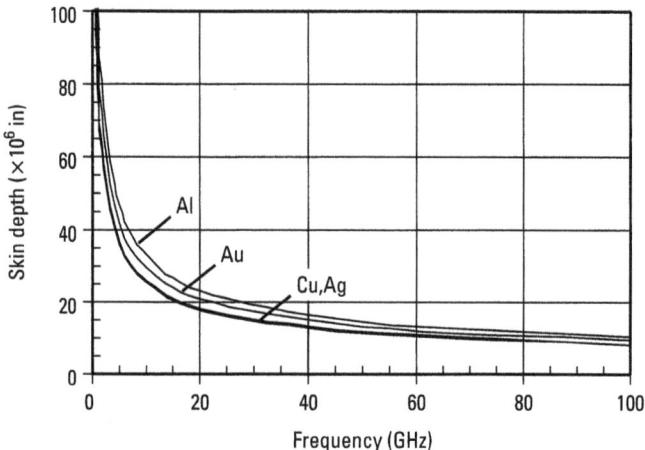

Figure 3.14 Skin depth of several common conductors.

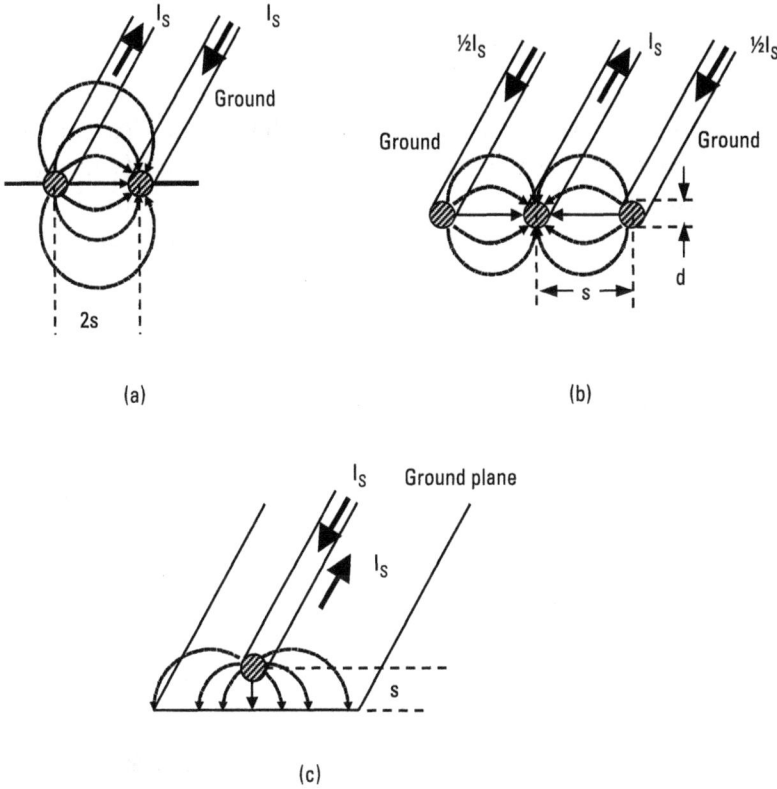

Figure 3.15 Common wire transmission lines and their TEM mode E-field distributions: (a) two-wire transmission line; (b) three-wire line; and (c) single wire over ground plane.

impedance of these transmission lines. For the two-wire transmission line in Figure 3.15(a), we have [6]

$$Z_{2wire} = \left(276/\varepsilon_r^{\,1/2}\right)\log_{10}\left(4s/d\right) \tag{3.13}$$

where $2s$ is the spacing between the centers of the wires, d is their diameter, and ε_r is the dielectric constant of the surrounding medium. For the three-wire line of Figure 3.15(b), the characteristic impedance is given by [6]

$$Z_{3wire} = \left(207/\varepsilon_r^{\,1/2}\right)\log_{10}\left(1.59s/d\right) \tag{3.14}$$

Figure 3.15(c) shows a single wire above a ground plane. If the distance above the ground plane is defined as s, then the impedance is approximately half that of the two-wire line:

$$Z_{wog} = \left(138/\varepsilon_r^{1/2}\right)\log_{10}\left(4s/d\right) \qquad (3.15)$$

Figure 3.15(c) is a simplified model of a bond wire that connects two microwave circuits such as the *microwave monolithic integrated circuit* (MMIC) and printed circuit board transmission line shown in Figure 3.16. Both components typically are mounted on a ground plane, with the interconnecting bond wire and the ground plane forming a nonuniform transmission line. The bond wire connection is shaped in an arch to provide mechanical strain relief, so the varying distance to the ground plane causes the characteristic impedance to vary significantly along its length. We use (3.15) to plot in Figure 3.17 the characteristic impedance of wires with different diameters as a function of ground plane spacing. The 0.001-inch (0.025 mm) diameter wire, in common use, has a characteristic impedance of 200 to 300 ohms when placed over a ground plane, causing a significant mismatch to a 50-ohm circuit. Increasing the wire diameter

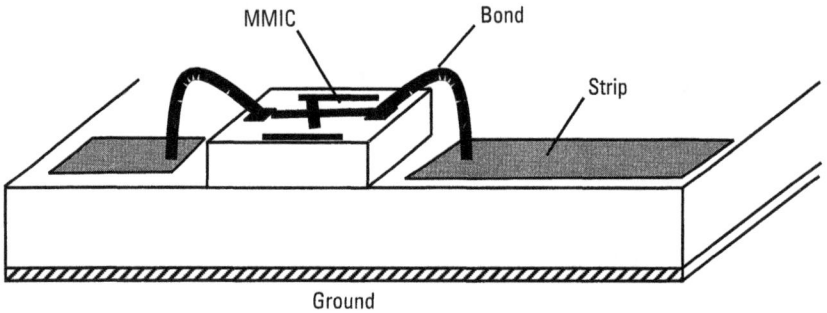

Figure 3.16 The gold bond wire used to connect an integrated circuit to a printed circuit board strip conductor is a wire transmission line over a ground plane.

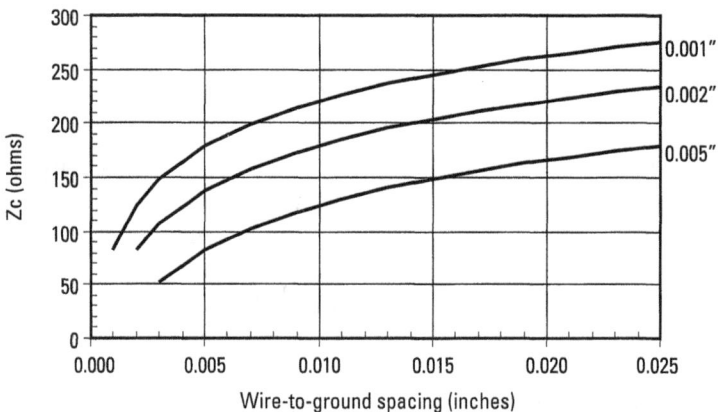

Figure 3.17 Characteristic impedance of 0.001-inch (0.025 mm), 0.002-inch (0.051 mm), and 0.005-inch (0.13 mm) diameter wire over ground versus ground plane spacing.

reduces the characteristic impedance and the mismatch. For the best match, microwave circuits occasionally are interconnected with 0.005-inch (0.13 mm) wide metal ribbon.

The previous discussion demonstrates that bringing a ground plane near an open wire transmission line changes its field profile and characteristic impedance dramatically. The current flowing in the ground plane flows in the opposite direction to that of the wire to satisfy the boundary condition that tangential electric field is zero on the ground plane surface. As we reduce the spacing between the ground plane and wire, the radiation from the currents tends to cancel, and we change the wire from a radiating antenna into a transmission line with a reasonably well-confined field. The coaxial transmission line is formed similarly by enveloping a single wire (the center conductor) with a concentric ground plane. In general, ground planes help confine the electromagnetic field and reduce radiation from current-carrying, open conductors.

3.4 Waveguides

A waveguide is a guided wave structure formed from a single enclosed conductor. Figure 3.18 shows the most common type, the *rectangular waveguide*. When we solve the Helmholtz equation (2.34) for the electromagnetic field modes that can exist within the waveguide, we find that the TEM mode is not a solution [7]. Instead, an infinite number of *transverse electric* (TE) and *transverse magnetic* (TM) modes may exist. These modes have transverse components of the electric and magnetic fields. The TE modes also have a propagational (z-directed) component of the magnetic field, and the TM modes have a propagational (z-directed) component of the electric field. The dominant mode for standard sized rectangular waveguide having width a greater than height b is the TE_{10} mode, which has the field components:

$$H_z = e^{-\gamma_{10}z} \cos(\pi x/a) \tag{3.16}$$

$$H_x = \gamma_{10}(\pi/a) e^{-\gamma_{10}z} \sin(\pi x/a) \tag{3.17}$$

$$E_y = -Z_{TE10} H_x \tag{3.18}$$

where $\gamma_{10} = \left[(\pi/a)^2 - k^2\right]^{\frac{1}{2}}$ is the propagation constant; $Z_{TE10} = jk\eta/\gamma_{10}$ is the mode's wave impedance; and $\eta = (\mu_0\mu_r/\varepsilon_r\varepsilon_0)^{\frac{1}{2}}$ is the *intrinsic impedance* of the dielectric filling the waveguide. The sinusoidal electric field profile is shown at the input to the waveguide in Figure 3.18. Propagation can occur only at frequencies greater than the cutoff frequency. The cutoff frequency for the TE_{10} mode is the frequency at which γ_{10} becomes equal to zero:

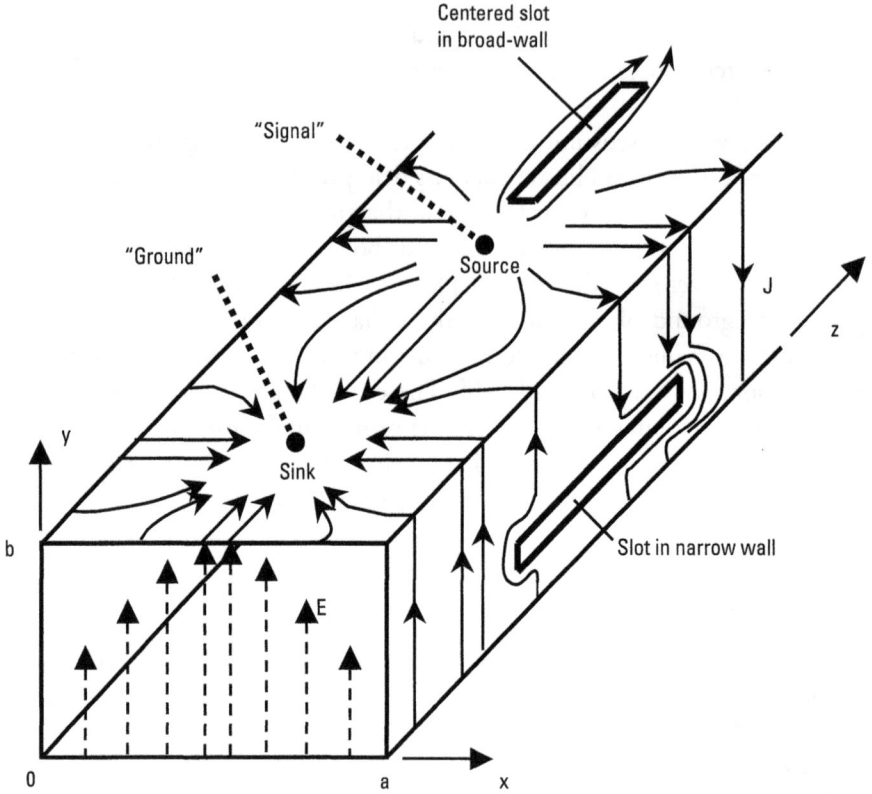

Figure 3.18 Rectangular waveguide showing dominant mode electric field **E** and surface current flow **J**.

$$f_{c10} = c/2a \qquad (3.19)$$

As compared to the TEM mode, the TE_{10} mode has a wave impedance that varies with frequency: it is infinite at the cutoff frequency and approaches the intrinsic impedance for $k \gg \pi/a$. The cutoff frequency for the next higher order mode, typically the TE_{20} mode, is twice f_{c10}. Consequently, rectangular waveguide has a single-mode bandwidth of 2 to 1. In practice, the bandwidth is restricted to about 1.5 to 1. For a propagating mode, the *guide wavelength* (the wavelength inside the waveguide) varies nonlinearly with frequency, and it is different for each mode. For the dominant mode, it is

$$\lambda_{g10} = j2\pi \Big/ \gamma_{10} = 2\pi \Big/ \left[\left(k^2 - (\pi/a)^2 \right) \right]^{1/2} \qquad (3.20)$$

The behavior of rectangular waveguide is described perfectly well by the electromagnetic field inside. But, as we know, the sources of the field are oscillating currents. Now, the walls of rectangular waveguide can always conduct current in one direction towards a load, and a *separate* conductor can provide a return path to ground. However, in this case we are just using the waveguide operating below cutoff as a wire, so the field inside must be zero. When an electromagnetic wave propagates within the waveguide, we discover that both signal and ground currents exist on the *same* conductor. We know that the surface current on a conductor is related to the magnetic field at the waveguide wall by (2.19), so we can determine the surface current for the TE_{10} mode from (3.16) and (3.17). The dominant mode surface current at one instant of time is plotted in Figure 3.18. On the upper broad-wall of the waveguide, sinks and sources of current are spaced half a guide wavelength apart. Since they move along with the propagating electromagnetic wave, half a cycle in time later, the sinks and sources will have exchanged places. Thus, for waveguides the selection of signal and ground current is completely arbitrarily, but it is clear that both exist and that they are flowing on the *same* conductor.

When we fabricate waveguide-based components, we want to use low loss metals and avoid discontinuities that block the current flow. Waveguide often is machined from aluminum for lightweight or brass if weight is not important. Copper is used also for its thermal properties. To prevent corrosion, the copper is nickel and gold plated. As with coaxial line, the gold plating thickness should be several skin depths to prevent the electrons from flowing in the more resistive nickel.

Cuts or slots in a waveguide component's walls can severely affect its performance depending on how they interrupt the flow of current. Slots are often intentionally cut into waveguide to form junctions or radiators. Two such slots, one in the broad-wall and the other in the narrow wall, are shown in Figure 3.18. If the broad-wall slot is thin ($< 0.1a$) and centered within the waveguide wall (at $x = a/2$) as in the figure, the current will flow around the slot with negligible change in insertion phase, and the slot will neither radiate nor reflect an incident TE_{10} mode. On the other hand, the same longitudinal slot cut into the narrow wall of the waveguide will force the current to make a significant detour (see Figure 3.18) to get to the other side. Even for a very narrow slot, the phase difference in the current on the two sides of the slot is sufficient to excite an electric field within the slot. When a wave propagating down the waveguide excites such a slot, it will radiate and cause the wave to lose energy. As an example, Figure 3.19 shows data from numerical simulations of a section of WR-15 waveguide with slots installed as in Figure 3.18. Figure 3.19(a) plots return loss versus frequency at the input to the waveguide for the centered broad-wall slot. As the slot width is increased, the electric field strengthens within the slot, and the return loss degrades. The return loss is very good (better than 30 dB) for slots widths less than $0.1a$. In Figure 3.19(b), we plot the insertion loss of the section of waveguide with a narrow wall longitudinal slot that is only 0.001 inch

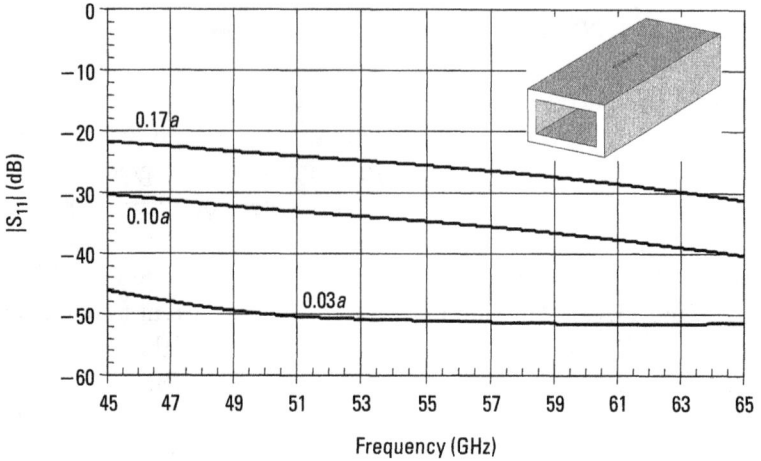

(a)

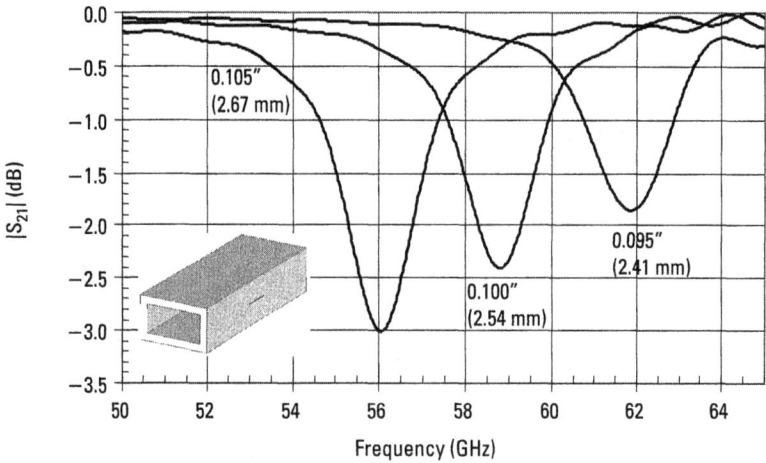

(b)

Figure 3.19 WR-15 [0.148 × 0.074 inch (3.76 × 1.88 mm)] rectangular waveguide with longitudinal slots. (a) $|S_{11}|$ versus slot width for 0.100 inch (2.5 mm) long, centered broad wall slot; and (b) $|S_{21}|$ versus slot length for 0.001 inch (0.025 mm) wide, centered narrow wall slot narrow wall.

(0.025 mm) wide (0.007a). Such a slot will radiate optimally when its length is about one-half of a free-space wavelength, 0.1 inch (0.25 mm) at 60 GHz. As the data show, the waveguide insertion loss is 2 to 3 dB at the slot's resonant frequency, which is very sensitive to slot length.

The behavior of slots in waveguide instructs us in the rudiments of waveguide component construction. Often, components such as filters are fabricated using a *split-block* assembly method in which the waveguide is machined from two pieces of metal that are aligned with precision pins and holes and screwed together. In splitting the waveguide, we are making a slot, and we want to minimize the interruption of current flow. Two ways to split the waveguide are along the middle of the broad-wall in the plane of the electric field, as in Figure 3.20(a), and through the narrow wall, along the plane of the magnetic field, as in Figure 3.20(b). A broad-wall split is nearly lossless when the waveguide pieces are screwed together. In general, we should avoid narrow wall splits because even the thinnest crack can radiate if it is a half wavelength or larger. In situations where narrow wall splits cannot be avoided, screws should be placed less than half a wavelength apart at the highest frequency of operation. Based on the data in Figure 3.19(b), the maximum separation should be no more than 0.4 wavelengths at the highest frequency of operation.

Besides rectangular waveguide, circular waveguide, shown in Figure 3.21, is quite common. The cutoff frequency of the dominant TE_{11} mode shown in the figure is given by the formula [8]

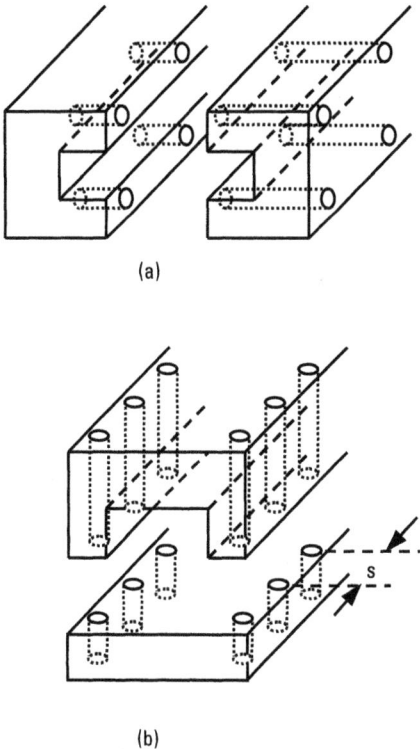

(a)

(b)

Figure 3.20 Split-block waveguide component construction. To minimize current flow interruption, (a) E-plane split is preferred to (b) H-plane split, for which the separation between holes s should be no more than 0.4λ.

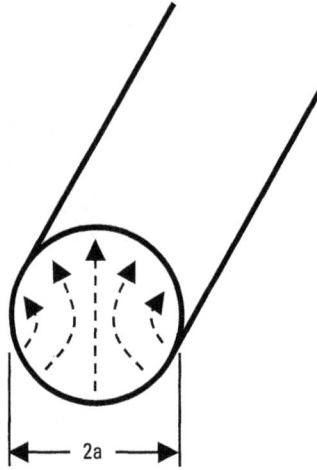

Figure 3.21 Circular waveguide and dominant TE$_{11}$ mode electric field distribution.

$$f_{c11} = 0.29304(c/a) \tag{3.21}$$

The dominant mode bandwidth is determined by the cutoff frequency of the next higher-order mode, the TM$_{01}$ mode, which is given by

$$f_{c01} = 0.38275(c/a)$$

3.5 Planar Transmission Lines

By far the most commonly used transmission lines are the planar types, some of which are shown in Figure 3.22. They can be constructed precisely using low-cost printed circuit board materials and processes. A number of these open, multiconductor transmission lines comprise a solid dielectric substrate having one or two layers of metallization, with the signal and ground currents flowing on separate conductors. The *parallel-plate waveguide* (PPWG) in Figure 3.22(a) propagates the TEM mode, and so does the *stripline* [Figure 3.22(c)]. For *microstrip* [Figure 3.22(b)], *coplanar waveguide* (CPW) in Figure 3.22(d) and slot line [Figure 3.22(e)], the electromagnetic field exists in two different dielectrics, the substrate and air, so the dominant mode is *quasi-TEM*. This mode has a 0-Hz cutoff frequency, and for most purposes is very much like the TEM mode, but its characteristic impedance and propagation constant slowly change with increasing frequency.

In general, for a given conductor separation (or substrate thickness) *h*, the characteristic impedance for PPWG, microstrip, and stripline decreases with

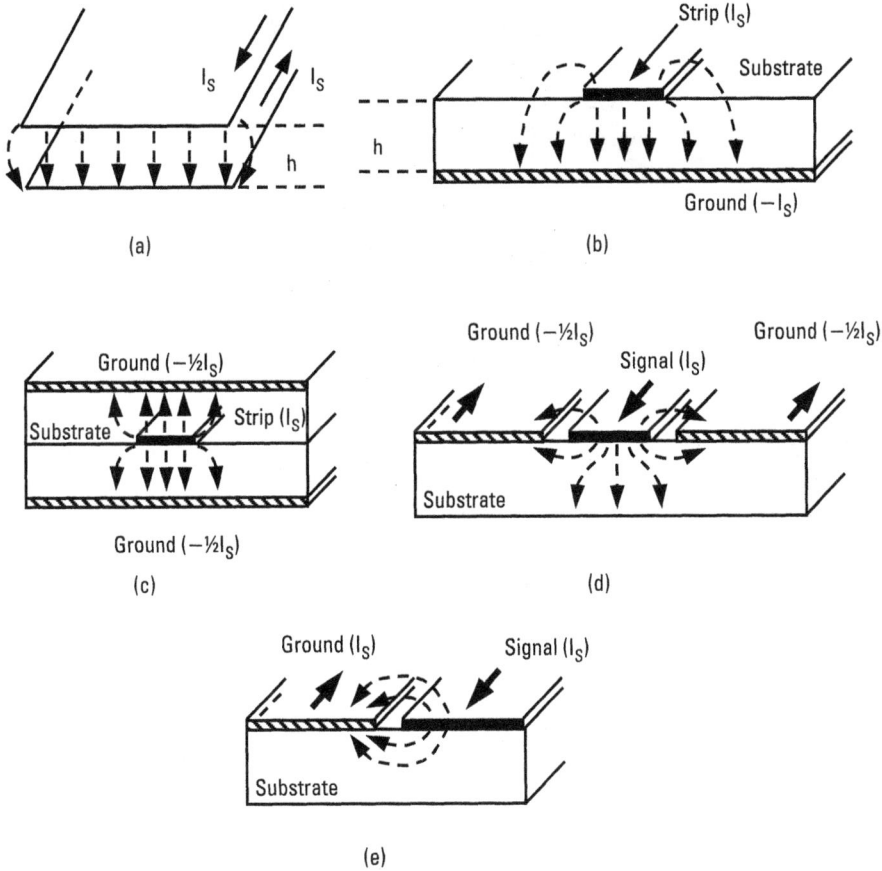

Figure 3.22 Common, multiconductor, planar transmission lines and their TEM mode E-field distributions: (a) parallel-plate, (b) microstrip, (c) stripline, (d) coplanar waveguide, and (e) slot line.

increasing line width and increasing dielectric constant [9, 10]. For CPW the characteristic impedance increases as the gap between the conductors or the center conductor (signal) line width decreases. Because both the signal and ground currents of CPW and slot line flow on the same layer, the substrate thickness is unimportant as long as it is much larger than the gap between the signal and the ground conductor. If there is a ground plane on the bottom of the substrate, then as the substrate thickness approaches the gap width, the electric field configuration will start to resemble that of microstrip [Figure 3.22(b)].

3.5.1 Microstrip

Microstrip [Figure 3.22(b)] is popular because it is well suited to applications requiring surface mount components, which attach directly to the top of the substrate. In simple terms, microstrip is the printed circuit version of a wire over a ground plane, and thus it tends to radiate as the spacing between the ground plane and the strip increases. A substrate thickness of a few percent of a wavelength or less minimizes radiation without forcing the strip width to be too narrow. The bandwidth of microstrip is also limited by coupling between the quasi-TEM mode and lowest order surface-wave spurious mode. This coupling becomes significant at a frequency given by [11]

$$f_T \cong \frac{150}{\pi h} \sqrt{\frac{2}{\varepsilon_r - 1}} \tan^{-1} \varepsilon_r \qquad (3.22)$$

where f_T is in gigahertz, ε_r is the dielectric constant of the substrate, and h, the thickness of the substrate, is in millimeters.

Microstrip's primary advantages of low cost and compact size are offset by its tendency to be more lossy than coaxial line and waveguide. One way to lower the loss of microstrip is to suspend it as in Figure 3.23(a). That is, we print the

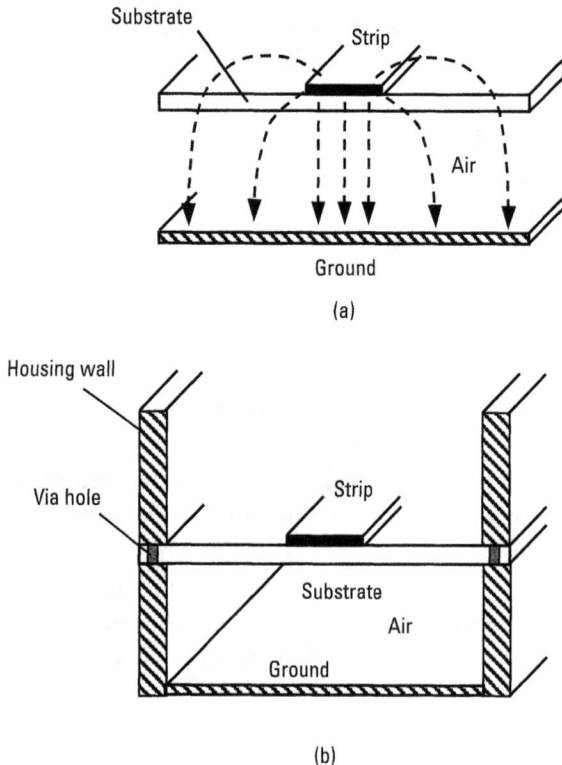

Figure 3.23 (a) Suspended microstrip transmission line. (b) Suspending microstrip line in a metal housing.

microstrip conductors on the top surface of a thin dielectric substrate and remove the ground plane metallization on its bottom surface. We then suspend the substrate some distance above a ground plane. The air in the intervening space between the bottom of the substrate and the ground plane contains the bulk of the electromagnetic field. The insertion loss of the microstrip is reduced for two reasons. Most importantly, air essentially has no dielectric loss compared to standard circuit board substrates. In addition, the width of the microstrip line increases because of the lower effective dielectric constant. Wider lines have lower current density, and thus, lower ohmic loss. However, suspending microstrip often means that the separation between the signal and ground paths increases, and so does the microstrip's tendency to radiate, particularly at discontinuities such as corners. Consequently, suspended microstrip mostly is used only up to a few gigahertz.

Microstrip circuits are often mounted in housings to isolate them from other devices. Passive circuits such as filters that require only a single layer of microstrip can be grounded by directly attaching the microstrip substrate ground plane to the housing base. Any coaxial connectors will be grounded to the housing also, thus insuring a continuous ground path for RF current. Transitions between coaxial connectors and microstrip are discussed at length in Chapter 4.

Suspended microstrip circuits are somewhat harder to ground to housings. Figure 3.23(b) shows one approach in which the substrate is held in place by its housing's upper and lower walls. Plated (via) holes drilled through the substrate provide a ground path that joins the upper and lower walls. In housings containing more than one such suspended substrate circuit, the via holes also suppress electromagnetic coupling between the circuits. One or more rows of vias should be drilled down the length of the circuit, spaced close enough to provide high isolation. A single row with at least eight and preferably more vias per wavelength is recommended. We examine isolation vias in detail in Section 5.5.2.

3.5.2 Stripline

Although microstrip requires only a single layer of dielectric, metallized on both sides, stripline [see Figure 3.22(c)] is often required for multilayer circuit boards because it can be routed between the layers. Grounding stripline requires some care. If the top and bottom ground planes are not at the same potential, a parallel-plate mode like that shown in Figure 3.22(a) can propagate between them. If excited, this mode will not remain confined to the region near the strip, but will be able to propagate wherever the two ground planes exist. In Figure 3.24, the parallel-plate mode is suppressed with metallized via holes connecting the two ground planes. The vias should be placed closely; a spacing s of one-eighth of a wavelength in the dielectric is recommended to prevent an appreciable potential

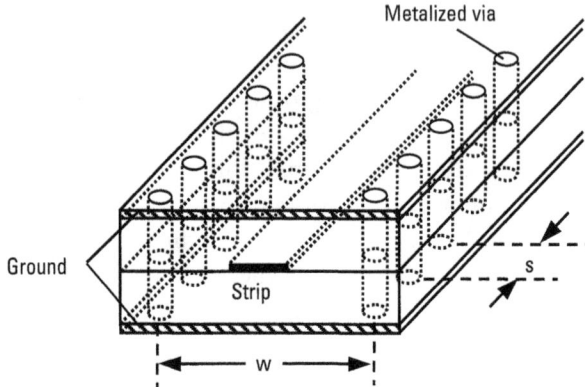

Figure 3.24 Stripline ground planes are maintained at the same voltage potential with closes spaced metallized via holes.

difference from forming between the ground planes. In addition, such vias form a cage around the strip, in essence making it a crude coaxial line. If this cage is to suppress leakage and coupling to other transmission lines that might be nearby, even closer spacing should be used. When the vias are placed too close to the edge of the strip, they can perturb its characteristic impedance. The via separation w should be a minimum of three strip widths, and five is preferable. However, if the via separation is too great, a pseudo rectangular waveguide TE_{10} mode can be excited (see Section 3.4). This mode has a cutoff frequency given by (3.19) as $c/2w$, where c is the speed of light in the dielectric. Thus, at the highest frequency of operation, f_{max}, the via separation w should be less than $c/2f_{max}$.

3.5.3 Coplanar Waveguide and Slot Line

CPW [Figure 3.22(d)] and slot line [Figure 3.22(e)] are alternatives to microstrip and stripline that place both the signal and ground currents on the same layer. They are the printed circuit analogs of the three-wire and two-wire transmission lines. Of the two transmission lines, CPW is used much more frequently in printed circuit boards. Like stripline, CPW has two ground planes, which must be maintained at the same potential to prevent unwanted modes from propagating. If the grounds are at different potentials, the CPW mode [see Figure 3.22(d)] will become uneven, with a higher field in one gap than the other. In Figure 3.25(a), bond wires are used to connect the ground planes and prevent this situation. The wires should be spaced one-quarter wavelength apart or less.

Figure 3.25(b) shows a diagram of grounded coplanar waveguide (GCPW), which is used on printed circuit boards as an alternative to microstrip line. For this application, the gap s between the strip and ground is usually more

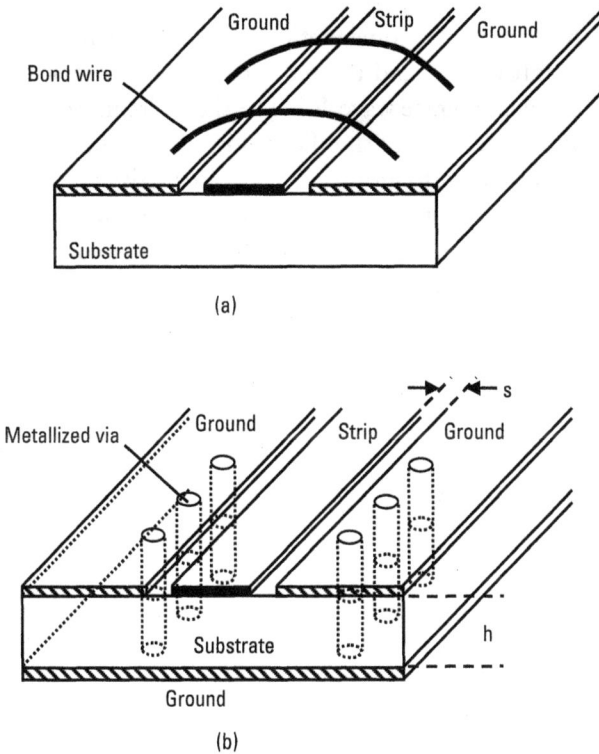

Figure 3.25 (a) Coplanar waveguide ground planes are maintained at the same voltage potential with bond wires or metal ribbons. (b) Grounded coplanar waveguide with ground vias for increased isolation.

than the thickness h of the substrate, so the GCPW field is concentrated between the strip and the substrate ground plane, and GCPW behaves like microstrip. With vias connecting the ground planes, GCPW is less prone to radiate and has higher isolation than microstrip.

3.6 DC Short Circuits and Via Holes

As we learned in Chapter 2, a short circuit connects two conductors formerly at different potentials and thus provides a path for current to flow from the higher potential conductor to the lower. A *DC short circuit* is a low impedance, physical joint (solder, conductive epoxy) that establishes the connection between conductors. In electronic circuit design, DC short circuits frequently are used to provide a direct path to ground for currents flowing on transmission lines and other components. A transmission line terminated in a short circuit to ground is terminated in a zero ohm load. Equation (3.5) tells us that the reflection coefficient of an incident electromagnetic wave is −1, meaning that all energy is reflected at the short circuit, and the reflected voltage wave V^-, is 180° out of phase with the incident voltage wave V^+. The short circuit carries the signal current to the ground plane without significant loss of potential.

At DC, a short circuit to ground is a very low resistance (ideally zero) connection, and an open circuit is a very high resistance (ideally infinite) connection. If again we invoke (3.5), we see that the reflection coefficient of an open circuit is +1, so the difference between open and short circuits to ground is simply 180° of phase. From (3.4), we can write the input impedance of a shorted transmission line ($Z_L = 0$) as

$$Z_{in} = jZ_0 \tan \beta z \tag{3.23}$$

For frequencies above 0 Hz, a microwave short circuit to ground has zero reactance only when z = 0 (i.e., when we are measuring the input impedance at the termination). As we move away from the termination (towards the source), the short circuit, as transformed by the transmission line, looks increasingly inductive. When $\beta z = \pi/2$, the transmission line is one-quarter of a wavelength long, and the input impedance becomes infinite.

Now let us consider Figure 3.26, which shows a microstrip line terminated in a metallized via hole connected to the ground plane. The use of via holes as short circuits for grounding planar transmission lines and surface-mounted components such as amplifiers is common practice in microwave printed circuit board design. If we apply the probes of a DC ohmmeter to the microstrip and ground plane, we measure the slight resistance of the metallization. However, we know that at increasing frequencies, the input impedance becomes inductive as the via hole's electrical length to the ground plane increases. Figure 3.26(b) shows the microwave equivalent circuit of a via hole, a series resistance and inductance. The oversized microstrip pad surrounding the via hole adds a small amount of capacitance, which we ignore here. Goldfarb and Pucel have derived frequency-dependent equations for both elements [12]:

$$R_{via}(f) = R_{DC}\left(1 + f/f_\delta\right)^{\frac{1}{2}}$$
$$R_{DC} = h \big/ \left\{\sigma\pi\left[r^2 - (r-t)^2\right]\right\} \tag{3.24}$$
$$f_\delta = \left(\pi\mu_0\sigma t^2\right)^{-1}$$

$$L_{via} = \frac{\mu_0}{2\pi}\left[h\ln\left(\frac{h + \sqrt{r^2 + h^2}}{r}\right) + \frac{3}{2}\left(r - \sqrt{r^2 + h^2}\right)\right] \tag{3.25}$$

(a)

(b)

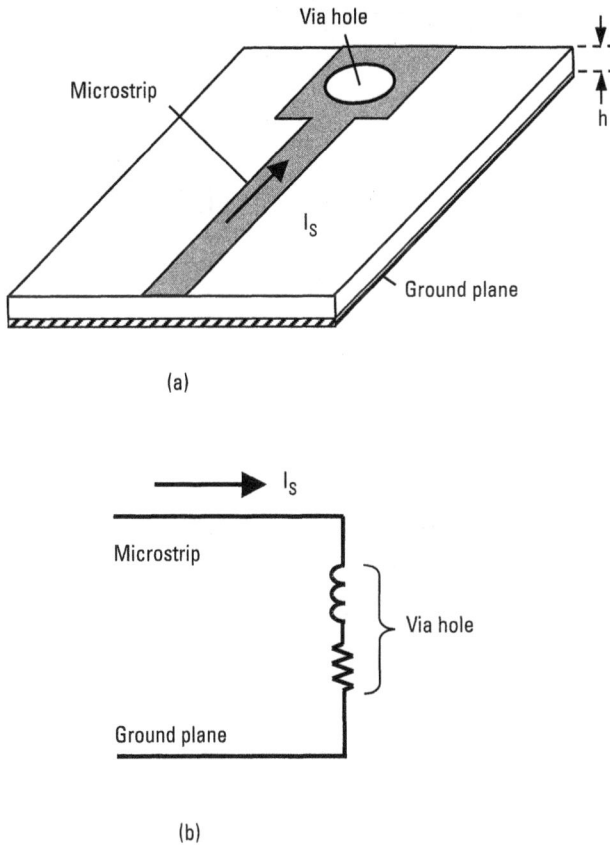

Figure 3.26 (a) Current flow down a microstrip line terminated in a via hole. (b) Equivalent circuit of microstrip line and via hole.

where r is the radius of the via hole, h is the substrate thickness, and t is the metallization thickness of the via walls. For a 0.02-inch (0.51 mm) diameter via plated with 0.001-inch (0.025 mm) copper in a 0.032-inch (0.81 mm) thick circuit board, the DC resistance is 0.007 ohm, and the inductance is 0.13 nH. At 10 GHz, the resistance is still low at 0.3 ohm, but the reactance is 8 ohms, hardly a perfect short circuit path to ground.

Figure 3.27 plots via hole inductance using (3.25) for different radii and substrate thicknesses. The plot shows that reducing the substrate thickness is a much more effective way of decreasing via hole inductance than increasing the radius of the via hole.

The work of Swanson suggests that (3.25) is a lower bound on via inductance [13]. Swanson used an electromagnetic field solver to compute equivalent inductances for the single and double-via grounded microstrip lines on an alumina substrate that are shown in Figure 3.28. Figure 3.29 plots the inductance

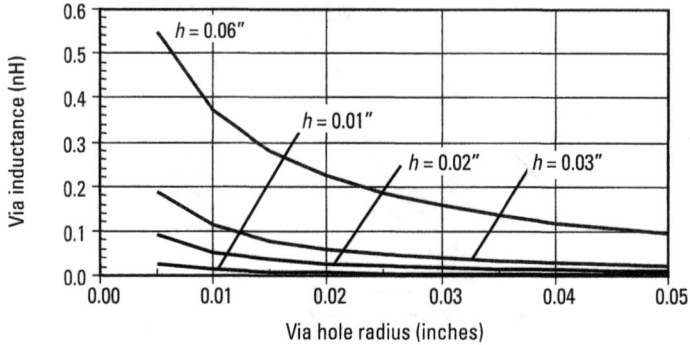

Figure 3.27 Via hole inductance versus radius as a function of substrate thickness *h*.

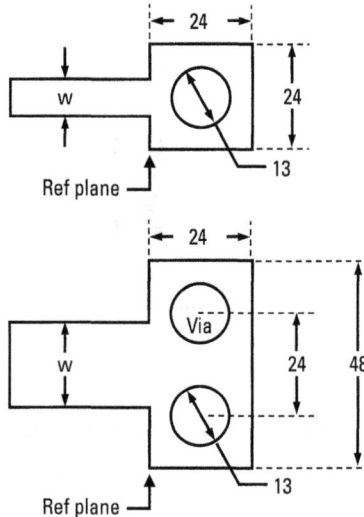

Figure 3.28 Single and double via holes terminating a microstrip line of width *w*. All dimensions are in mils (1 mil = 0.001 inch = 0.025 mm). (*From:* [13]. © 1992 IEEE.
Reprinted with permission.)

versus frequency of single vias in 0.015-inch (0.38 mm) and 0.025-inch (0.63
mm) thick substrates for several microstrip line widths. Evidently, the inductance of the termination is a function not only of the via hole parameters but
also of the microstrip line width. For the single via in a 0.015-inch (0.38 mm)
substrate, (3.25) predicts an inductance of 0.045 nH, about one-third of
Swanson's result. Figure 3.30 plots the inductance for the double-via terminations of Figure 3.28. The second via lowers the inductance to ground by about
one-third.

Swanson validated his via hole model and demonstrated the influence of
inductance in the ground path by designing an interdigitated filter shown in

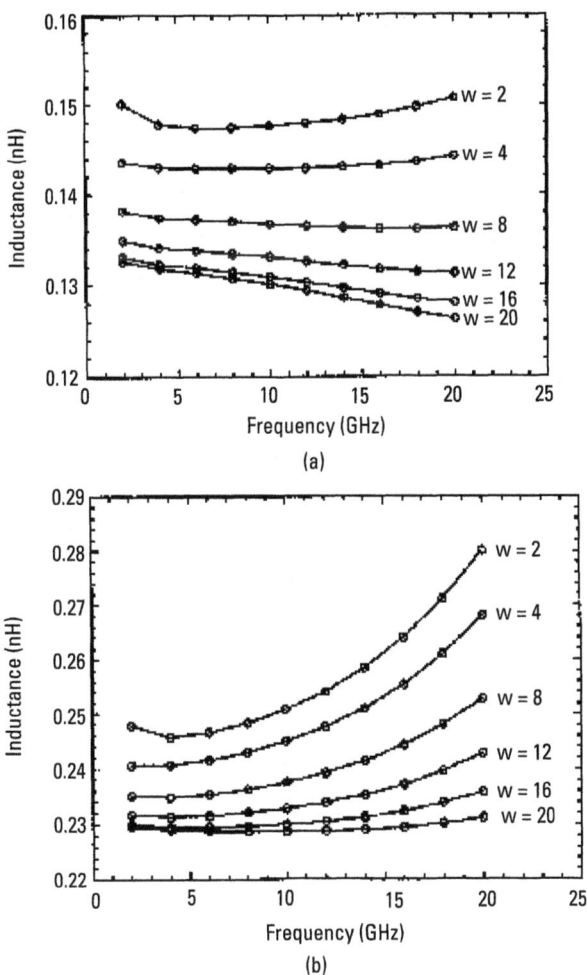

Figure 3.29 Inductance of a single via hole terminating a microstrip line of width *w* in mils: (a) 0.015 inch (0.38 mm) thick; and (b) 0.025 inch (0.63 mm) thick alumina substrate. $\varepsilon_r = 9.8$. (*From:* [13]. © 1992 IEEE. Reprinted with permission.)

Figure 3.31. Energy couples into the filter through the horizontal taps on either side of the structure. The designer selects the lengths and spacings of the coupled vertical lines to achieve the desired passband and out-of-band rejection. Each microstrip line is grounded in a double-via. Figure 3.32(a) compares the filter's measured response with the simulated response for ideal (zero-inductance) vias. The error in the prediction is about 4%. Figure 3.32(b) replots the simulated response, which includes Swanson's double-via inductance model; the agreement with measurement is nearly perfect.

Via holes generally are drilled completely through a circuit board, and the walls are plated with metal to a thickness of about one-thousandth of an inch

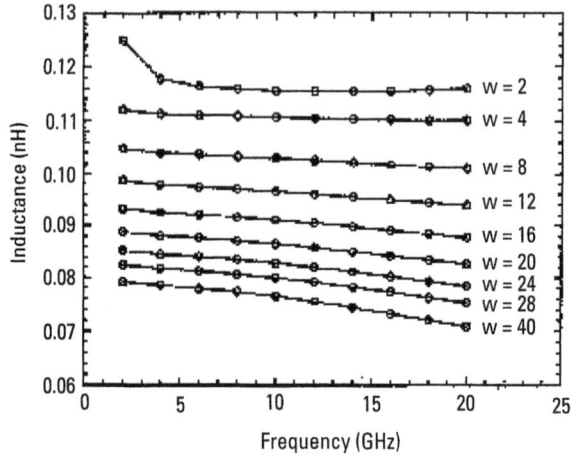

(a)

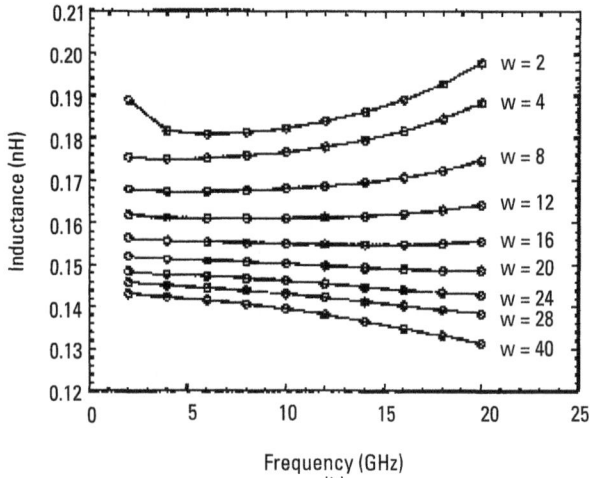

(b)

Figure 3.30 Inductance of a double via holes terminating a microstrip line of width *w* in mils: (a) 0.015 inch (0.38 mm) thick; and (b) 0.025 inch (0.63 mm) thick alumina substrate. $\varepsilon_r = 9.8$. (*From:* [13]. © 1992 IEEE. Reprinted with permission.)

(0.025 mm). Occasionally, the via holes are filled completely with metal. A filled via has slightly lower DC resistance than a plated via, but since AC current flows on the outside of a conductor, the primary advantage that a filled via has over a plated via is its lower thermal resistance. When active components such as amplifiers must be mounted on top of a printed circuit board, an array of filled vias under the active device provides a better thermal path than an array of plated vias.

In summary, a transmission line terminated in a DC short circuit to ground has a reflection coefficient of unity with a phase of 180°. However, as

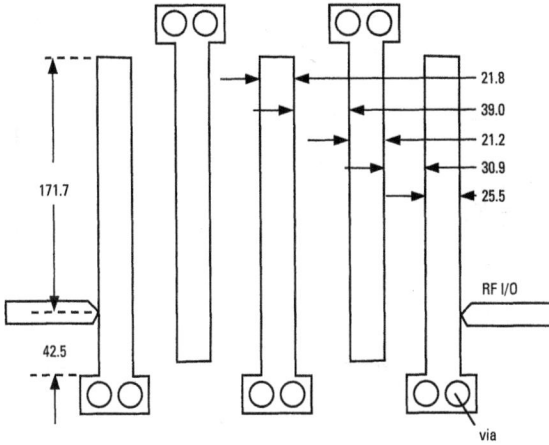

Figure 3.31 Microstrip interdigital filter on 0.015-inch (0.38 mm) thick alumina substrate. ε_r = 9.8. (All dimensions are in mils.) (*From:* [13]. © 1992 IEEE. Reprinted with permission.)

frequency increases, the inductance in the short circuit causes the phase to deviate from 180°. The amount of phase change increases with short circuit electrical length, and thus an imperfect DC short circuit has a bandwidth that is inversely proportional to its electrical length.

3.7 RF Short and Open Circuits

As frequency increases, particularly into the millimeter-wave bands, the reactance to ground of a DC short circuit such as a via hole becomes unacceptable. For example, a 0.1-nH via hole has 22 ohms of reactance at 35 GHz. As an alternative to a DC short to ground, we can use an *RF short circuit*, a quarter-wavelength transmission line terminated in an open circuit. Figure 3.33 shows both DC and RF short circuit terminations realized with microstrip. Unlike the ideal DC short circuit, which is a zero-ohm termination connecting the signal and ground conductors of a transmission line, a RF short circuit is a *noncontacting* termination. The transmission line signal and ground connections are purposely left unconnected so as to form an open circuit. No current flows to the ground conductor from the signal conductor, although current is induced in the ground plane. The incident and reflected voltage waves create a standing wave for which the voltage one-quarter wavelength back from the open circuit is zero volts as shown in Figure 3.34. From (3.4), the input impedance of a transmission line terminated in an open circuit is given by

$$Z_{in} = -jZ_0 \cot \beta z \tag{3.26}$$

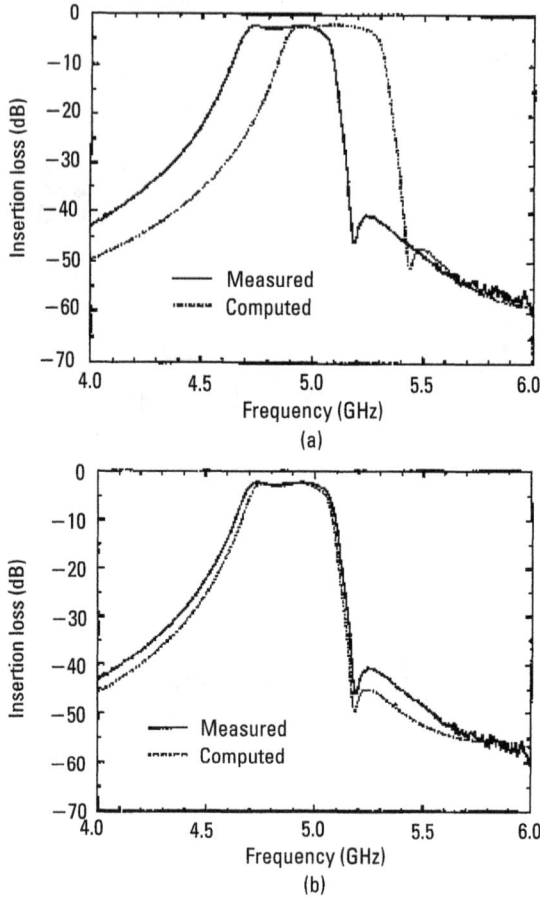

Figure 3.32 Measured and computed insertion loss of the interdigital filter: (a) without and (b) with double-via inductance. (*From:* [13]. © 1992 IEEE. Reprinted with permission.)

If $z = \lambda/4$, then $\beta z = \pi/2$, $\cot \beta z = 0$, and $Z_{in} = 0$. Unlike a DC short circuit, which is a perfect short at 0 Hz, the input impedance of a transmission line terminated in a RF short circuit is 0 ohms at the frequency for which the transmission line is one-quarter wavelength long. At other frequencies its reactance increases according to (3.26). At low frequencies near 0 Hz, the RF short circuit becomes an open circuit.

Microstrip open circuits are imperfect. An open-circuited transmission line stores charge at its open end. For planar transmission lines such as microstrip, a charge of equal amplitude and opposite polarity is induced in the ground plane. The pair of charges forms a fringing capacitance between the two conductors, causing the transmission line to behave as if it were longer. We shorten the transmission line's length to compensate for this *open-end effect*.

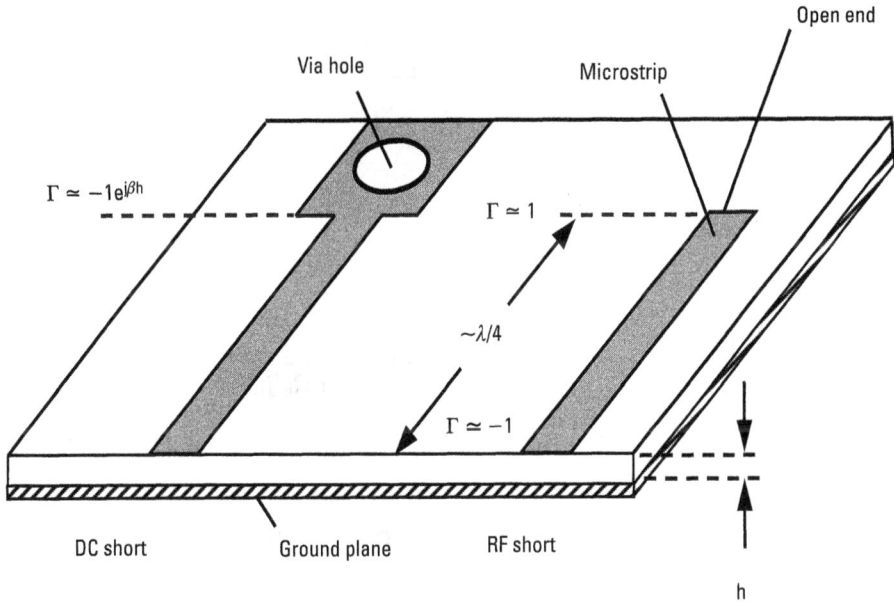

Figure 3.33 A via hole is a perfect short circuit to ground only at 0 Hz because it has physical length h, while an open-circuited line can be transformed into a perfect short circuit at any frequency.

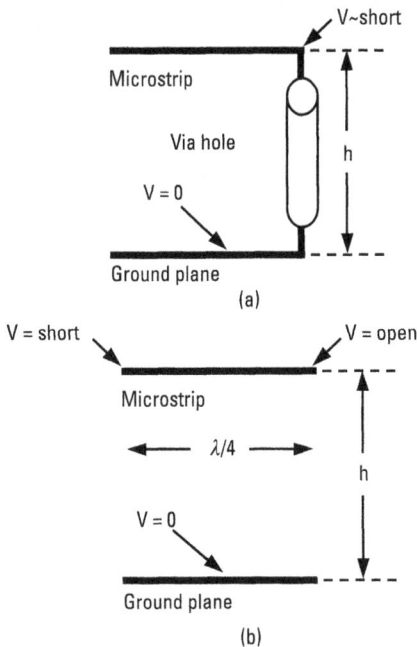

Figure 3.34 Equivalent circuit of (a) via hole to ground, and (b) RF short from quarter-wave open-circuited line.

An *RF open circuit* is obtained by terminating a quarter-wave length transmission line in a short circuit. Its input impedance is given by

$$Z_{in} = jZ_0 \tan \beta z \qquad (3.27)$$

with $z = \lambda/4$, $\beta z = \pi/2$, $\tan \beta z$ is infinite, and Z_{in} is infinite, an open circuit. Like the RF short circuit, the reactance of the RF open changes with frequencies, and at 0 Hz, the RF open circuit becomes a short circuit.

Applications of RF short and open circuits are widespread in microwave circuits. RF shorts often are required in amplifier and oscillator bias circuits to prevent microwave current from flowing in DC bias lines. At RF frequencies below 10 GHz, where the wavelength becomes quite long, instead of open-ended transmission lines, we may use more compact lumped capacitors to suppress unwanted RF as we describe in Section 3.8.2.

Other applications of RF shorts and opens include a diplexer, which is a T-junction terminated in two filters. A series-type waveguide diplexer cannot operate properly without a good RF short circuit at the input to the out-of-band filter. RF open circuits are useful as choke slots to shape the radiation pattern of antennas as described in Chapter 6. For planar circuits, RF short circuits are useful when via hole grounding is not possible. For instance, millimeter-wave circuits that require 50-ohm terminations may use an RF short circuit rather than a DC short circuit as shown schematically in Figure 3.35. In a narrow bandwidth, the quality of a termination employing a RF short circuit can be superior to a DC grounded termination. Figure 3.36 compares two 50-ohm loads used to match a transmission line having a characteristic impedance of 50 ohms. The DC load uses a 0.1-nH via to connect a 50-ohm resistor to ground, while the RF load uses a transmission line tuned to be a quarter-wavelength long at 20 GHz as its termination. Over a band of about 5 GHz, the return loss of the RF load exceeds that of the DC load. For wide bandwidth, a DC short circuit is preferable, but as frequency and the inductance in the ground path increase, an RF short circuit can be superior over a significant frequency band.

3.8 Printed Circuit Boards

Planar transmission lines are well suited for use in a multilayer microwave *printed circuit board* (PCB), which consists of one more layers of metallized dielectric laminated together and interconnected with via holes. Active and passive surface mount components are soldered or epoxied to the top layer. A microwave PCB often carries mixed signals, in that one or more of its layers conduct microwave signals, and the others conduct the analog and digital signals that bias and control the microwave devices. In this section we focus on the

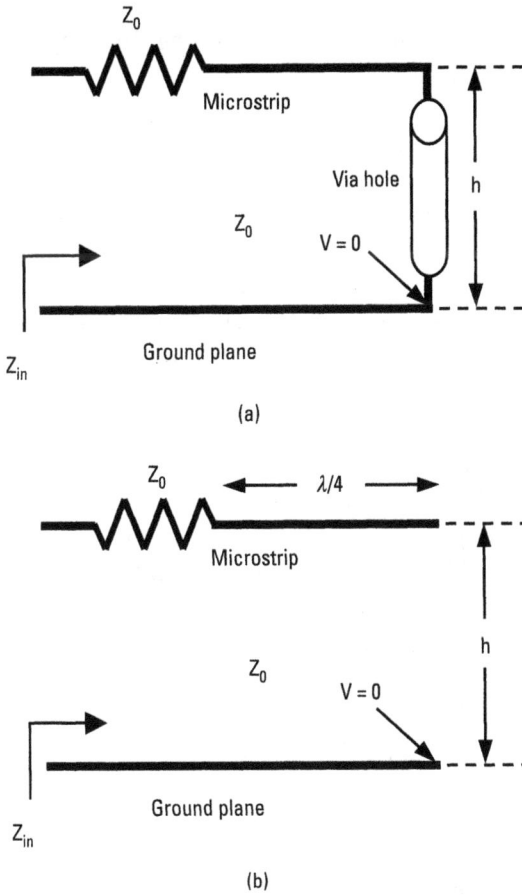

Figure 3.35 Equivalent circuit of 50-ohm loads using (a) via hole to ground, and (b) RF short from quarter-wave open-circuited line.

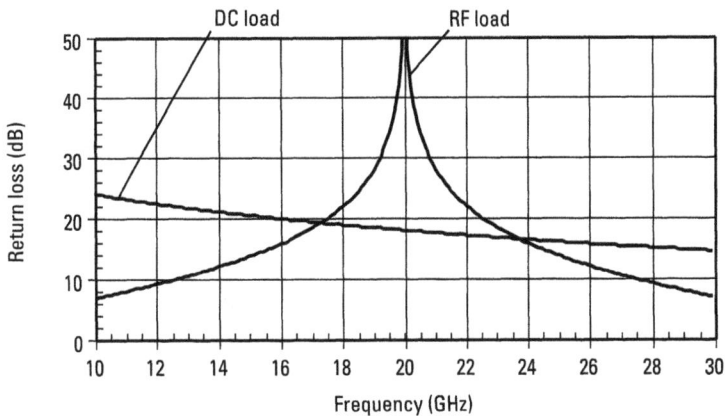

Figure 3.36 Return loss of a 50-ohm transmission line terminated in a 50-ohm resistor DC grounded through a 0.1-nH via hole, and RF grounded by a quarter-wave open-circuited line tuned at 20 GHz.

passive circuit aspects of circuit board grounding. In Chapter 5, we will discuss the grounding of active devices on circuit boards.

3.8.1 Layer Definition and Grounding

Figure 3.37 depicts a typical four-layer circuit board that carries microwave, DC, and digital signals [14]. The top conductor is the microwave signal layer, and it carries the signals between amplifiers, mixers, filters, and other microwave devices that might be used in a multifunction circuit board. For the optimum microwave performance and impedance control, a low loss material with a nearly frequency-independent permittivity and well-constrained thickness should be chosen as the top layer of dielectric material. The next layer down is the RF ground plane, ideally a solid conductor that carries the microstrip ground current and isolates the RF signals from the digital and DC signals beneath it. As the grounding scheme for a microwave PCB generally is multipoint, one should avoid using the RF ground plane as the digital and DC ground plane also. Otherwise, digital, DC, and RF ground currents will be coupled, and unwanted noise will be transferred to sensitive RF circuits such as detectors and oscillators. The third conductive layer can be used to route signal and ground currents for DC and digital circuitry. The bottom conductor may be used for additional routing of DC and digital signals or it may serve as the DC/digital ground plane.

RF, DC, and digital surface mount components are soldered to the top layer of the circuit board. Via holes (see Figure 3.37) are used to make the connections between these components and the other layers. Typically, via holes are drilled and metallized after the circuit board layers have been etched and

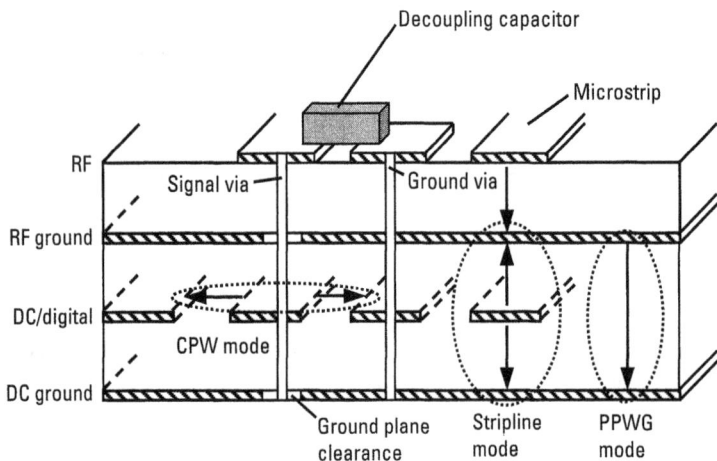

Figure 3.37 This four-layer circuit board with RF, digital, and DC conductors has the potential to propagate unwanted signals between the two ground planes.

laminated together, so they must pass through the entire circuit board. Metallization must be cleared on all ground planes where signal vias pass through to avoid unwanted short circuits. Similarly, vias going to ground cannot make contact with signal traces on interior layers.

Ideally, the microwave, digital, and DC ground currents return to ground on their respective ground planes and are completely uncoupled. However, the RF and DC ground layers are connected, generally by via holes. A via has some impedance; so there will be a voltage difference between the RF and DC ground layers. Most circuit boards have hundreds of vias, which create multiple parallel ground paths, all with nonzero impedance. We know from Chapter 1 that the RF current in each via will be coupled with the current in the other vias. RF current that finds its way between the RF and DC ground planes can excite a variety of unwanted microwave transmission line modes as shown in Figure 3.37. Energy in these modes can travel for some distance within the circuit board and couple through the vias back to the RF conductor layer. Such a bypassing transmission line can degrade the performance of active components. It can cause an amplifier to become unstable and oscillate (see Chapter 5) or degrade the isolation of a filter as shown in Figure 3.38. If all ground planes within a circuit board are interconnected using via holes such that the potential difference

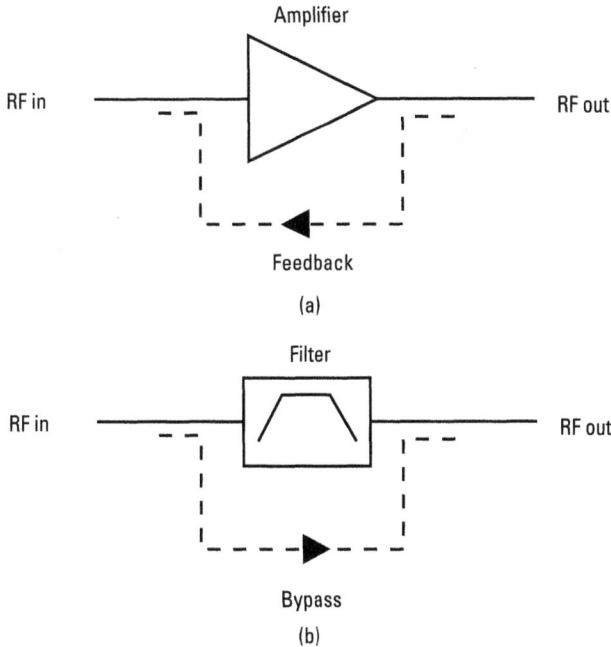

Figure 3.38 Poor grounding on a multilayer circuit board enables microwave signals to propagate in unintended layers: (a) feedback causes an amplifier to oscillate; and (b) filter isolation is degraded by a bypassing transmission line.

between them is made small, then very little energy can couple into parallel-plate modes. Further, one should place metal grounded with vias wherever possible on unused areas of each conductor layer, and connect all ground planes through the board with vias spaced as closely as practical (at least one-eighth of a wavelength). Patches of metal that cannot be grounded should be avoided as they can radiate as antennas. A line of grounded vias running around the perimeter of the circuit board or a plated board edge helps suppress board radiation. Signal traces should be surrounded as much as possible by grounded metal. Structures like stripline (see Figure 3.24) and grounded coplanar waveguide [see Figure 3.25(b)] make good choices if space permits.

As an example, Figure 3.39(a) shows the top layer of a two-layer millimeter-wave transceiver circuit board. Signals that propagate on this layer are microstrip and DC bias, and the bottom layer is the RF and DC ground plane. The ground plane is attached with conductive epoxy to the circuit board metal housing. To maximize isolation between the various parts of the circuit, a metal cover with machined channels shown in Figure 3.39(b) fits over the transmission lines and components on the top of the circuit board. The cover touches each of the grounded metal patches. Channelization of circuit boards is discussed further in Section 5.5.3.

3.8.2 Decoupling Methods

Vias to ground are not always sufficient to suppress unwanted modes. In Figure 3.37, stripline and CPW modes still may be excited between DC or digital signal conductors and the ground planes. In Figure 3.39, we do not want RF signals to propagate on the bias lines, and we do not want noise on the bias lines to modulate the output of the RF devices. However, we cannot simply use vias to ground the signal conductors without grounding the DC and digital signals too. A simple solution, shown in Figure 3.37, uses a *decoupling capacitor* to shunt any unwanted RF current to ground. The capacitor is mounted to the top layer of the circuit board, and vias are used to connect it between the ground plane and the desired trace on the DC/digital layer. The reactance of the capacitor, $(2\pi \times$ frequency $\times\ C)^{-1}$, increases as frequency or capacitance decreases. In effect, the capacitor short-circuits the CPW center conductor to ground at RF frequencies, but it is open-circuited at low frequencies, leaving the digital and bias currents unperturbed. In Figure 3.39, the decoupling capacitors short high frequency signals to ground while appearing as open circuits to the DC bias current. For example, a 0.1-μF capacitor is a good RF short circuit at 1 GHz, with a reactance of only 1.5 ohms, but at 10 MHz, its reactance is 160 ohms.

Decoupling capacitors are not effective in suppressing noise from digital and DC circuitry above 1 or 2 GHz. The typical surface mount capacitor has a series RLC circuit model, and a self-resonant frequency of $[2\pi(LC)^{\frac{1}{2}}]^{-1}$. Above

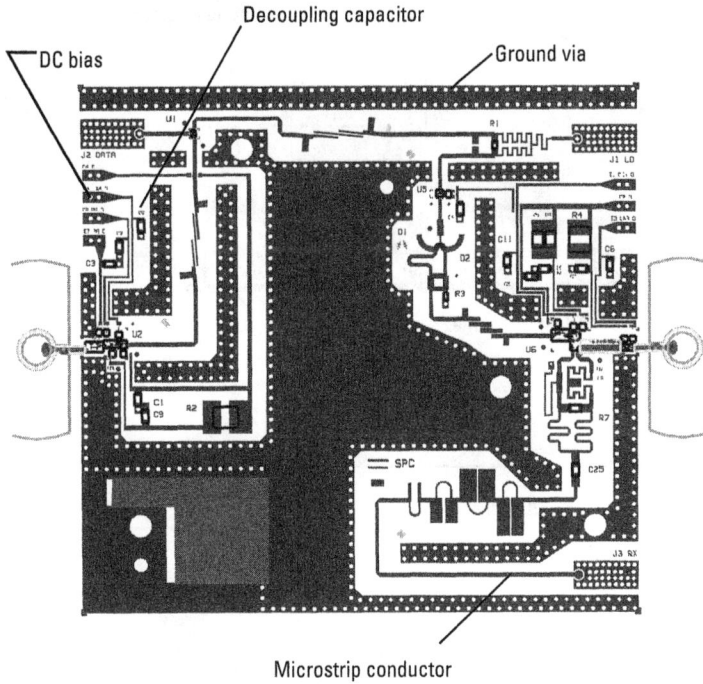

Decoupling capacitor

DC bias

Ground via

Microstrip conductor

(a)

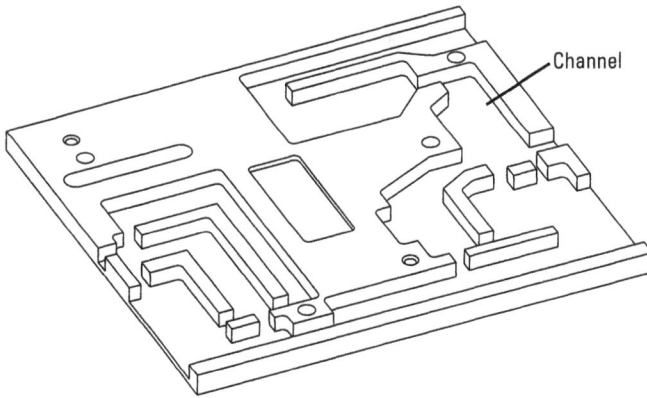

Channel

(b)

Figure 3.39 (a) Signal layer of two-layer millimeter-wave circuit board. (b) Circuit board metal cover channelizes RF transmission lines.

that frequency, the package's parasitic inductance tends to overwhelm the capacitance. As a higher frequency alternative, a number of workers have been using *electromagnetic band gap* (EBG) structures to suppress unwanted coupling between the RF and non-RF active components on a printed circuit board [15–20]. Figure 3.40(a) shows a parallel-plate mode propagating between the

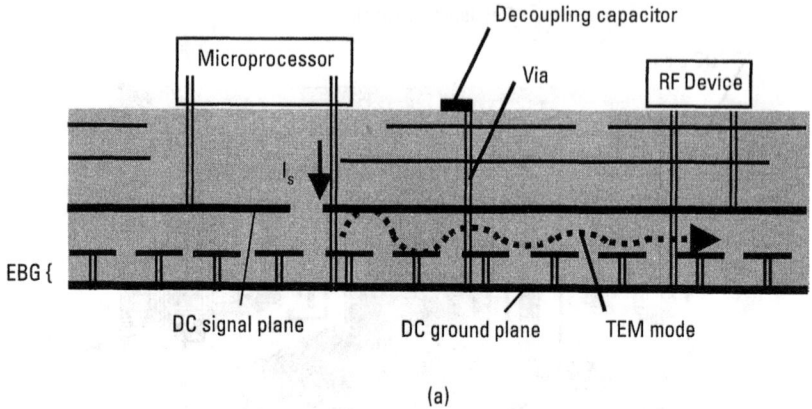

(a)

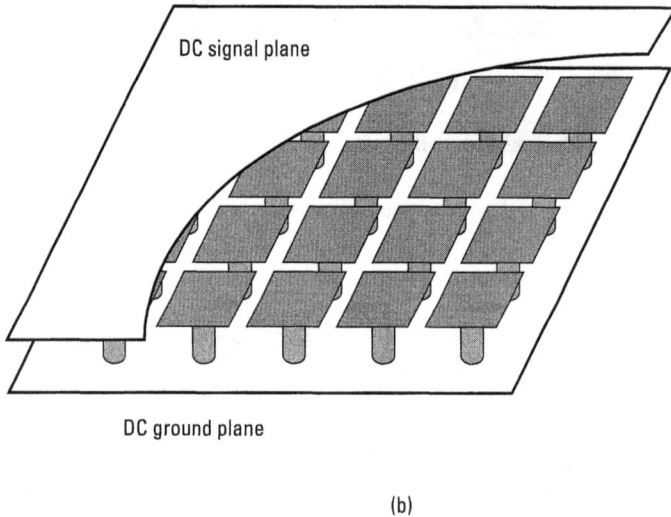

(b)

Figure 3.40 (a) An EBG structure can be placed between the signal and ground planes within a circuit board to suppress parallel-plate mode coupling between high frequency components; and (b) EBG construction. (*After:* [15].)

DC ground and signal planes in a printed circuit board. To prevent the mode from propagating, an EBG layer is inserted between the two planes. As shown in Figure 3.40(b), an EBG structure is an array of square, printed metal patches connected by vias to the ground plane. The EBG functions as a type of high impedance surface. In this case, its dimensions are selected to reject the parallel-plate waveguide mode over a stop band. The stop band moves higher in frequency with increasing patch size, and increasing the capacitance between the EBG and the power plane increases the stop bandwidth. Figure 3.41 shows some typical performance that can be achieved with a configuration like that in

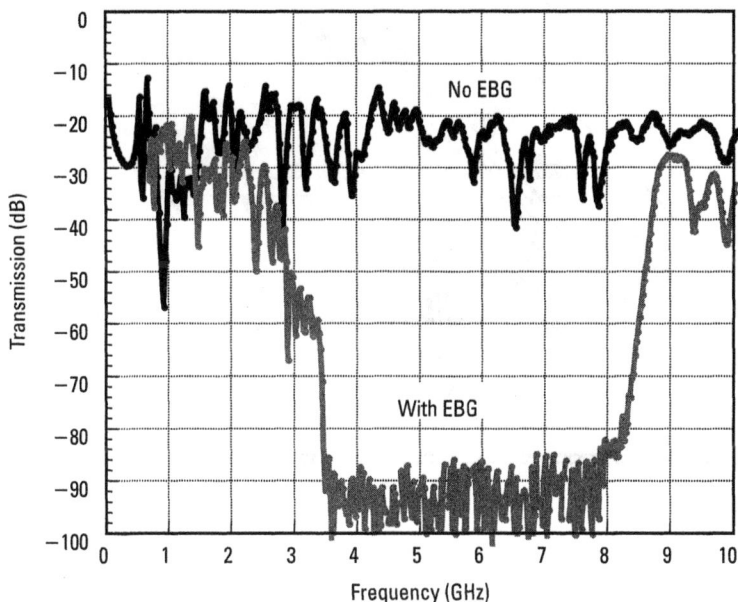

Figure 3.41 The electromagnetic band gap layer provides over 50-dB reduction in coupling within a printed circuit board over a 2:1 bandwidth. (*After:* [15].)

Figure 3.40. Over roughly the 4 to 8 GHz stop band, the EBG layer increases signal rejection by 50 to 60 dB.

Electrical engineers often combine RF and lower frequency analog circuitry on the same layers of a printed circuit board to reduce cost. To isolate the RF and analog circuitry on the same plane from the digital circuitry, they may split the ground plane with slots as shown in Figure 3.42. In the case of the straight slot in Figure 3.42(a), RF and analog circuitry is placed on one side of the straight slot, and the digital circuitry is placed on the other side. A DC-to-DC power converter circuit might be placed inside the boundary of the moat slot in Figure 3.42(b), while the lower power RF and analog circuitry might be placed outside. In either case, a narrow path of interconnecting metallization maintains the regions of the split ground plane at the same potential. If a microstrip line passes over the slot as in Figure 3.42, the ground plane will carry the current returning to the source. A slot interrupting the ground plane current can radiate and thus provide a mechanism to couple signals between the conductors above and below the ground plane.

Moran et al. have investigated ways to reduce coupling between microstrip lines and various types of ground plane slots using measurements and rigorous electromagnetic simulations [21]. To minimize the coupling between a microstrip line with a characteristic impedance that usually is 50 ohms and a slot, we want the slot to have a high reactance. Thus, we want to avoid sending

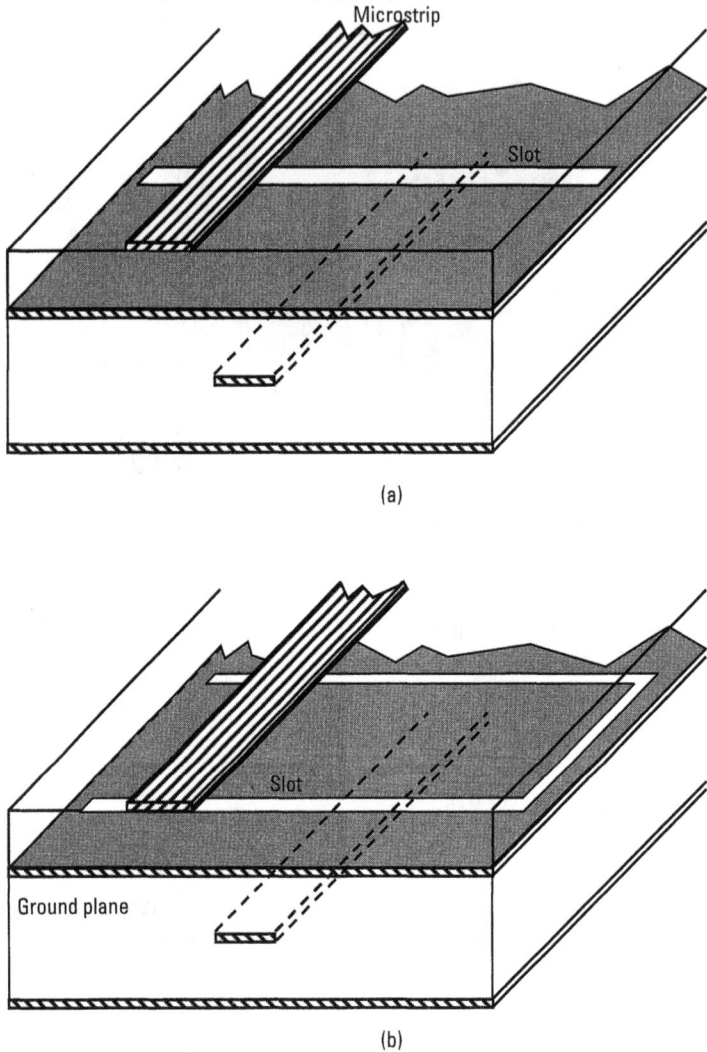

Figure 3.42 Microstrip with (a) straight slot and (b) moat-type slot in the ground plane.
(*After:* [21].)

signals through the microstrip line at the resonant frequencies of the slot, where its reactance goes to zero. At such frequencies, energy will couple strongly from the microstrip line to the slot and be radiated to the other side of the ground plane. But even well away from resonant frequencies, Moran's team found that the slots in Figure 3.42 radiate when traversed by a microstrip line. One way to prevent the slot from radiating is to bring the ground current up to the signal layer by transitioning to a different transmission line type such as coplanar waveguide, as shown in Figure 3.43. However, coplanar transmission line requires

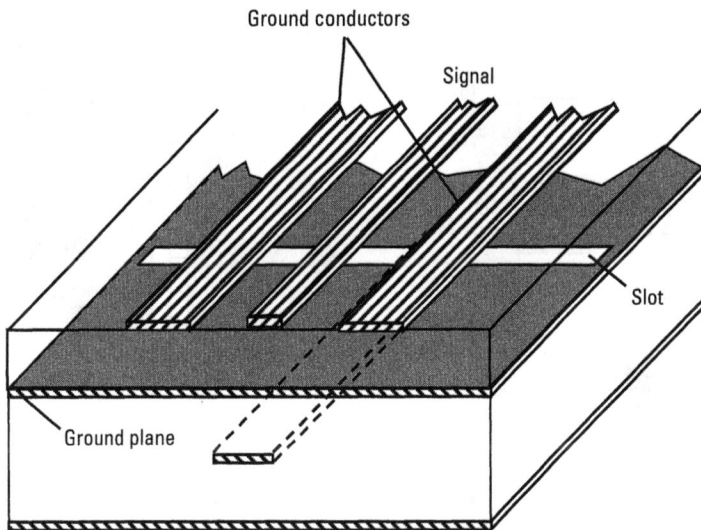

Figure 3.43 A CPW transmission line crosses over a ground plane slot. The slot does not radiate because all ground current flows in the CPW ground conductors, and none flows in the ground plane.

significantly more circuit board space than a microstrip. Another approach is to modify the slot geometry. Figure 3.44(a) shows an RF choke slot. The lengths and widths of the sections of the choke slot affect both the resonant frequencies and the frequency band over which the choke slot exhibits reduced radiation compared to the straight and moat slots. Like the coplanar transmission line, the choke slot fills circuit board space that could accommodate circuit elements. A more effective slot is the corrugated slot shown in Figure 3.44(b). A mesh of very narrow conductors divides the slot into many smaller, highly reactive slots. The mesh conductors are sufficiently narrow to maintain isolation between the split ground plane regions. Moran showed that the corrugated slot has no resonances over a wide bandwidth, and compared to the choke slot, it requires less circuit board space.

Even when unwanted transmission modes within the circuit board are suppressed by proper grounding, surface mount components such as filters still may suffer degraded performance if they are poorly attached to the circuit board. Most surface mount packaged filters have their input and output ports at opposite ends of the package, which serves as the ground reference for both ports. The mounting pins of filter packages such as that shown in Figure 3.45(a) have inductance, and they enable the package and printed circuit board grounding system to form bypassing transmission paths. The common ground plane under the filter can inductively couple the current flowing in the input and output pins. This coupling limits the out-of-band rejection for pin-mounted filters

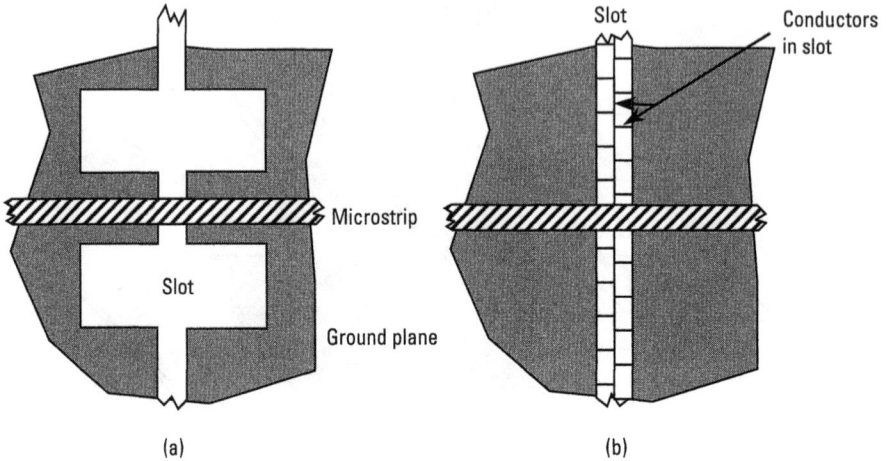

Figure 3.44 Microstrip with (a) choke slot and (b) corrugated slot in the ground plane. (*After:* [21].)

to about 60 dB [22]. Figure 3.45(b) shows that improper mounting of a pin-less package to the circuit board can leave a small gap under the filter package, which can form a bypassing parallel-plate transmission line that degrades the filter's

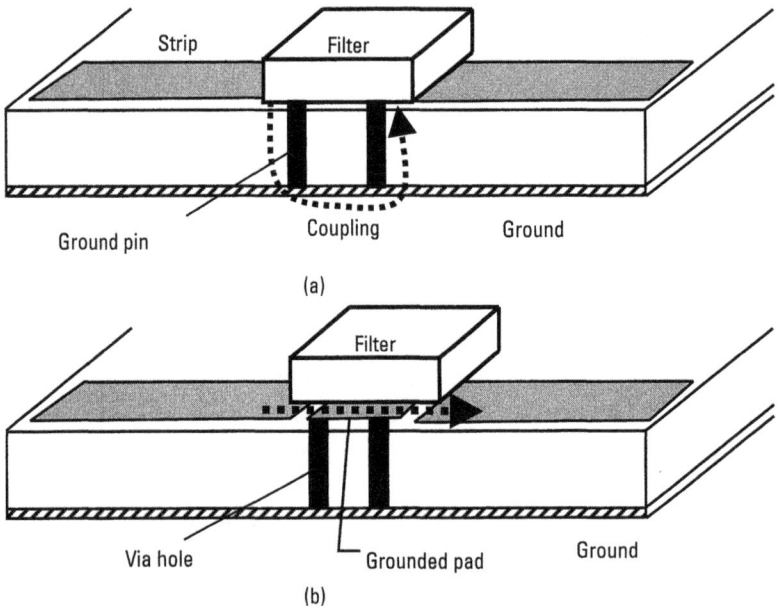

Figure 3.45 Filter rejection is degraded by (a) inductive coupling between ground pins and the circuit board ground, and (b) a bypassing transmission line formed between the filter package and mounting pad.

rejection. This transmission line can be shorted out by grounding the filter package thoroughly at its input and output ports. If out-of-band rejection exceeding 60 dB is required, coaxial input and output ports may also be used. Because they are not grounded through the circuit board, coupling is reduced significantly.

For miniature multilayer ceramic filters like the one shown in Figure 3.46(a), grounding can be a challenge. With dimensions that are comparable to those of lumped element, chip resistors and capacitors, these filters can routinely achieve 60 dB of rejection outside their passband. Since their input and output ports are separated by less than 0.1 inch (2.5 mm), proper grounding is essential to suppress bypassing leakage. Figure 3.46(b) shows a typical printed circuit board RF layout [23]. A grounded isolation bridge separates the RF input and output transmission lines, and the circuit board ground pad is connected to the RF ground plane with several vias placed directly under the filter package.

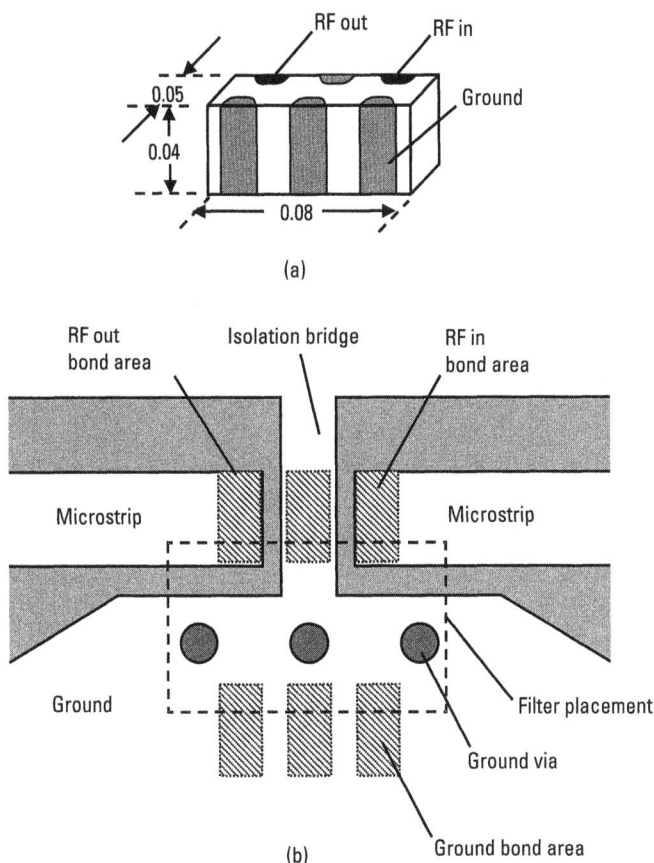

(a)

(b)

Figure 3.46 (a) A typical, miniature, multilayer ceramic bandpass filter package (dimensions in inches). (b) Circuit board mounting and grounding layout. (*After:* [23].)

References

[1] Cheng, D. K., *Fields and Wave Electromagnetics*, Reading, MA: Addison-Wesley, 1983, pp. 353–356.

[2] Collin, R. E., *Foundations for Microwave Engineering*, New York: McGraw-Hill, 1992, pp. 248–257.

[3] Ramo, S., J. R. Whinnery, and T. Van Duzer, *Fields and Waves in Communications Electronics*, 2nd ed., New York: John Wiley & Sons, 1985, pp. 428–430.

[4] Jordan, E. C., *Electromagnetic Waves and Radiating Systems*, Upper Saddle River, NJ: Prentice Hall, 1950, pp. 164–165.

[5] Collin, R. E., *Foundations for Microwave Engineering*, New York: McGraw-Hill, 1992, pp. 254–257.

[6] Jasik, H., *Antenna Engineering Handbook*, New York: McGraw-Hill, 1984, Chapter 42, p. 9.

[7] Balanis, C. A., *Advanced Engineering Electromagnetics*, New York: John Wiley & Sons, 1989, pp. 352–387.

[8] Balanis, C. A., *Advanced Engineering Electromagnetics*, New York: John Wiley & Sons, 1989, Section 9.2.

[9] Cheng, D. K., *Fields and Wave Electromagnetics*, Reading, MA: Addison-Wesley, 1983, pp. 372–375.

[10] Balanis, C. A., *Advanced Engineering Electromagnetics*, New York: John Wiley & Sons, 1989, pp. 444–455.

[11] Bahl, I., and P. Bhartia, *Microwave Solid State Circuit Design*, 2nd ed., New York: John Wiley & Sons, 2003, p. 38.

[12] Goldfarb, M. E., and R. A. Pucel, "Modeling Via Hole Grounds in Microstrip," *IEEE Microwave Guided Wave Lett.*, Vol. 1, No. 6, June 1991, pp. 135–137.

[13] Swanson, Jr., D. G., "Grounding Microstrip Lines with Via Holes," *IEEE Trans. on Microwave Theory and Techniques*, Vol. 40, No. 8, August 1992, pp. 1719–1721.

[14] Bremer, R., T. Chavers, and Z. Yu, "Power Supply and Ground Design for WiFi Transceiver," *RF Design*, November 2004, pp. 16–22.

[15] Rogers, S., et al., "Noise Reduction in Digital/RF Daughter Card with Electromagnetic Bandgap Layers," unpublished presentation, *IEEE 13th Topical Meeting on Electrical Performance of Electronic Packaging*, October 2004.

[16] Rogers, S., "Electromagnetic-Bandgap Layers for Broad-Band Suppression of TEM Modes in Power Planes," *IEEE Trans. on Microwave Theory and Techniques*, Vol. 53, August 2005, pp. 2495–2505.

[17] Abhari, R., and G. V. Eleftheriades, "Suppression of the Parallel-Plate Noise in High Speed Circuits Using a Metallic Electromagnetic Band-Gap Structure," *2002 IEEE MTT-S International Symposium*, Vol. 51, June 2002, pp. 493–496.

[18] Abhari, R. and G. V. Eleftheriades, "Metallo-Dielectric Electromagnetic Bandgap Structures for Suppression and Isolation of the Parallel-Plate Noise in High-Speed Circuits," *IEEE Trans. on Microwave Theory and Techniques*, Vol. 51, June 2003, pp. 1629–1639.

[19] Kamgaing, T., and O. M. Ramahi, "High-Impedance Electromagnetic Surfaces for Parallel-Plate Mode Suppression in High Speed Digital Systems," *IEEE 11th Topical Meeting on Electrical Performance of Electronic Packaging*, October 2002, pp. 279–282.

[20]   Kamgaing, T., and O. M. Ramahi, "A Novel Power Plane with Integrated Simultaneous Switching Noise Mitigation Capability Using High Impedance Surface," *IEEE Microwave and Wireless Components Letters*, Vol. 13, January 2003, pp. 21–23.

[21]   Moran, T. E., et al., "Methods to Reduce Radiation from Split Ground Planes in RF and Mixed Signal Packaging Structures," *IEEE Trans. on Advanced Packaging*, Vol. 25, No. 3, August 2002, pp. 409–416.

[22]   Lark Engineering Company, "The Better the Ground...the Better the Performance," http://www.larkeng.com/GENINFO/Ground.htm.

[23]   TOKO America, Inc., "Recommended Solder Pad Layout LTF2012B-F Series," unpublished application note.

4

Transmission Line Transitions

Transmission line transitions are required in nearly every microwave subsystem and component. A transition is an interconnection between two different transmission lines that possesses low insertion loss and high return loss. These characteristics can be achieved only through careful matching of the impedances and electromagnetic fields of the two transmission lines. The designs of the signal and ground current paths through a transition are also critical. For a transition to function properly, these paths must often be continuous, in close proximity to suppress radiation, and as short and closely matched in length as possible. As we study transitions between planar, coaxial, and waveguide transmission lines in this chapter, we focus on the relationship between the design of signal and ground current paths and transition performance.

4.1 Fundamentals and Applications

We already know that a coaxial transmission line possesses wide bandwidth and provides high isolation from external signals, so it is well suited for interconnecting microwave modules and systems that cannot tolerate interference. From Section 3.2 we know that to avoid the propagation of higher order modes as frequency increases, we must reduce the circumference of coaxial line. However, as the inner and outer conductor diameters decrease, the current density increases and so does the resistive loss. Consequently, at millimeter-wave frequencies (above 28 GHz), waveguide, which has the lowest insertion loss of conductor-based transmission lines, often replaces coaxial line as the energy transporter of choice. In general, though, the weight, size, and cost of coaxial line and waveguide preclude their use within most microwave modules. Instead, to route electromagnetic signals inside modules between components such as oscillators, amplifiers, and filters, most engineers use low cost planar transmission line-based circuit boards. This use of planar transmission lines inside modules and coaxial line and waveguide outside means that most modules require a transition at every RF interface. Figure 4.1 shows two such transitions, one from microstrip to coaxial line, and the second from microstrip to circular waveguide.

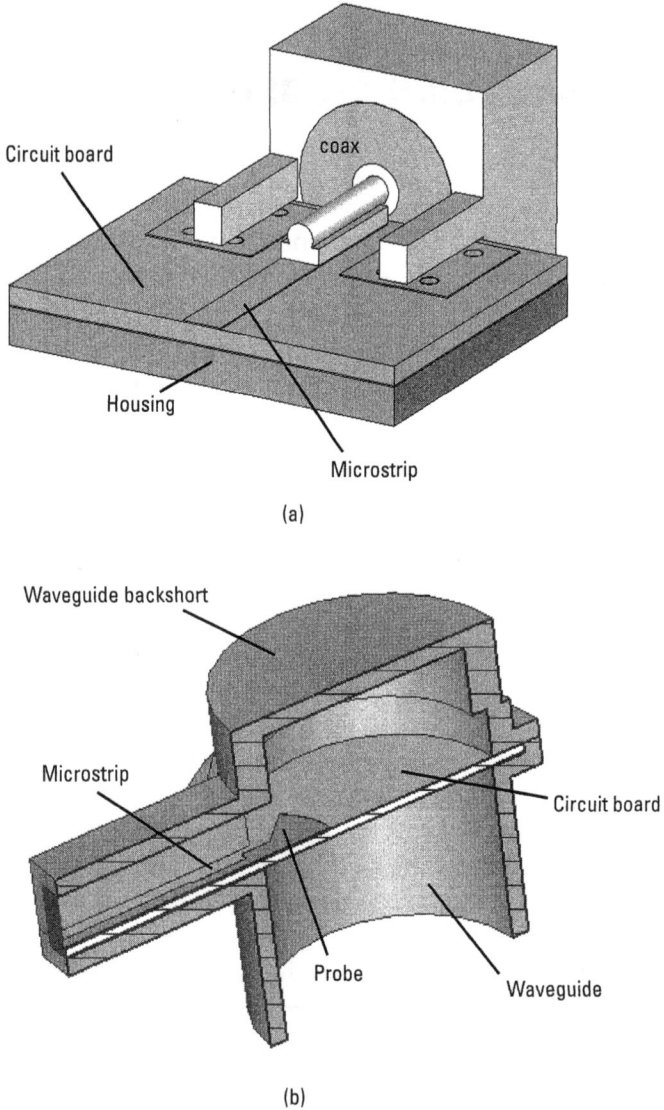

(a)

(b)

Figure 4.1 Transmission line transitions: (a) coaxial line to microstrip; and (b) circular waveguide to microstrip.

Physically, a transition is a nonuniform structure that connects two different transmission lines. The transition enables the cross-sectional geometry of the signal and ground conductors to change from that of one transmission line to that of the other. The signal and ground currents may take separate, divergent, or even discontinuous paths as they flow through the transition. Most transitions work best when we keep these paths continuous, short, parallel, close together, and closely matched in length to minimize mismatch and radiative losses.

A well-designed transition converts the transverse field configuration and characteristic impedance of one transmission line to that of another transmission line over a desired frequency band of operation while maintaining low insertion loss and high input return loss. Insertion loss—the amount of power exiting the transition—expressed as a fraction of the input level, should be less than 0.25 dB. Return loss—the amount of power reflected at the transition—expressed as a fraction of the input level, should be at least 15 dB. Additionally, a good transition should be easy to fabricate, mechanically robust, and insensitive to ambient temperature variations.

A transition's primary operational limitation is its frequency bandwidth, which, to a first order, depends on the modal characteristics of the transmission lines involved.[1] We can categorize transmission lines into two groups by their conductor configuration, as shown in Figure 4.2: multiconductor structures that propagate the TEM mode, and single-conductor waveguides with a dominant mode (typically TE) having a cutoff frequency above 0 Hz. In a typical transition, the transmission line with the narrowest dominant mode bandwidth limits the maximum bandwidth of the transition.

A transition between coaxial line and microstrip such as that in Figure 4.1(a) can have extremely wide bandwidth because both coaxial line and microstrip are TEM transmission lines with frequency-independent impedances and 0-Hz dominant mode cutoff frequencies. In principle, the upper end of the transition's bandwidth is limited only by how small we can make the cross-sectional dimensions of these transmission lines. On the other hand, if we replace the coaxial transmission line with a circular waveguide as in Figure 4.1(b), then the cutoff frequency of the circular waveguide's dominant TE_{11} mode sets the lower limit of the transition's passband. The next higher order mode, the TM_{01}, sets the upper limit.

Transitions between TEM transmission lines often employ a direct physical connection between their respective signal and ground conductors so that current can flow even at 0 Hz. On the other hand, if one of the transmission lines is a waveguide, a direct connection between the ground of the microstrip

---

1. Although transitions involving higher order modes occasionally find use in systems, in this chapter we will assume that all transmission lines are propagating their dominant mode only.

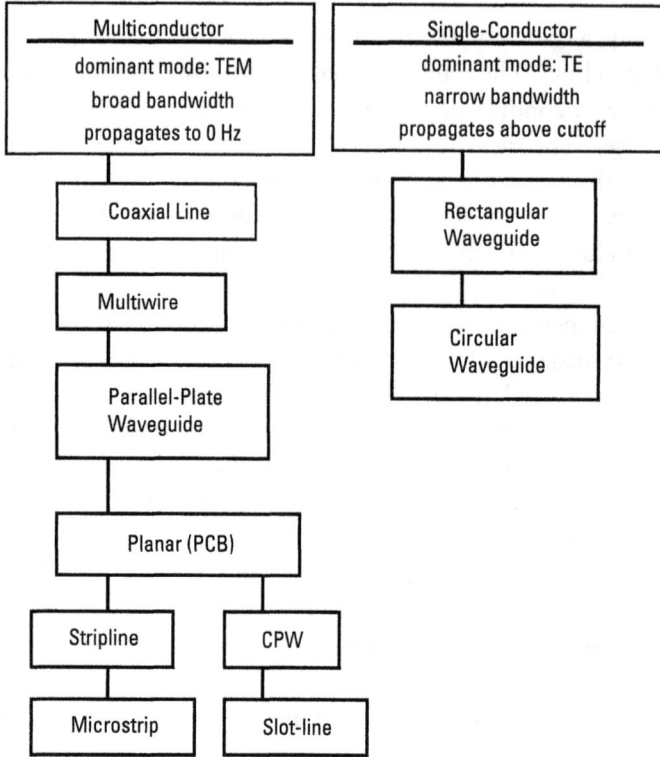

Figure 4.2 The maximum bandwidth of a transition depends on the bandwidths of the transmission lines involved.

and the waveguide wall typically is made, while the strip signal current may couple to the currents flowing on the waveguide walls by radiation from a strip probe as shown in Figure 4.1(b).

If we view a transition as an ideal enclosed two-port circuit, then conservation of energy says that energy entering one port will exit the other port. However, many transitions are not ideal two-ports; a transition that is not fully enclosed by conductors [see Figure 4.1(a)] may be well matched and yet exhibit unexpectedly high insertion loss from radiation. Excess loss from radiation often occurs in transitions when the ground path is electrically long or discontinuous or when the ground and signal paths become too widely separated. One should always be aware that even currents flowing through well-designed transitions exhibit some level of radiation, which can be the source of unwanted interference. For example, high power signals passing through a coax-to-microstrip transition may interfere with sensitive receiver circuitry in the vicinity. Such a transition may require grounded shielding to reduce its interaction with the receiver circuitry. We devote the remainder of this chapter to study the importance of grounding in transition design.

4.2 Coaxial Line to Microstrip Transitions

Transitions between coaxial line and microstrip are commonly used on printed circuit boards. The TEM mode characteristic impedances of coaxial line and microstrip often are the same (typically 50 ohms), so the design effort simplifies to matching their field configurations and designing the signal and ground current paths. The two most popular configurations are *edge-mounted*, in which the coaxial transmission line is attached to the microstrip line at the edge of a circuit board, and *vertical-mounted*, in which the coaxial line center conductor passes through the circuit board and intersects the microstrip orthogonally from below.

4.2.1 Edge-Mounted Transitions

At first glance, the field configuration of coaxial line seems quite different than that of microstrip (see Figure 4.3), but both transmission lines confine the electromagnetic field between two conductors. The coaxial field is uniformly distributed around the center conductor while the microstrip field is concentrated in the substrate under the strip. A transition of this type exhibiting nearly optimum performance is Eisenhart's edge-launch design, shown in Figure 4.3 [1]. As the coaxial line's center conductor gradually slopes down towards the circuit board, the electromagnetic field becomes concentrated below the center conductor like that in the microstrip line. In addition, the signal and ground current paths are well matched, being in close proximity, nearly parallel, and differing in length only slightly (see Figure 4.3). Eisenhart's transition achieves very high performance: he has demonstrated greater than 25-dB return loss up to 18 GHz. Such a transition is ideal for testing prototype microstrip circuits. However, its significant length and relative high cost make it undesirable as a transition for mass-produced microwave circuit boards. Besides, most microwave circuit interfaces do not require the bandwidth of the Eisenhart transition.

Figure 4.4(a) shows an *edge launch transition* between a microstrip on a two-layer circuit board and a subminiature version A (SMA) coaxial line. The board-mounted coaxial connector includes a protruding center conductor that is soldered to the microstrip. Ground pins attached to the edges of the coaxial connector make the ground contact to metallized pads on top of the circuit board and to either side of the microstrip. Since the microstrip ground plane lies between this circuit board's two dielectric layers, we use via holes to carry the coaxial line's ground current through the circuit board. The vias on each pad should be spaced no more than one quarter-wavelength apart to keep the pads from radiating. Figure 4.4(b) shows the surface currents flowing on the transition, with a pair of signal and ground current paths delineated. While the signal current follows a path between the center conductor and the microstrip that

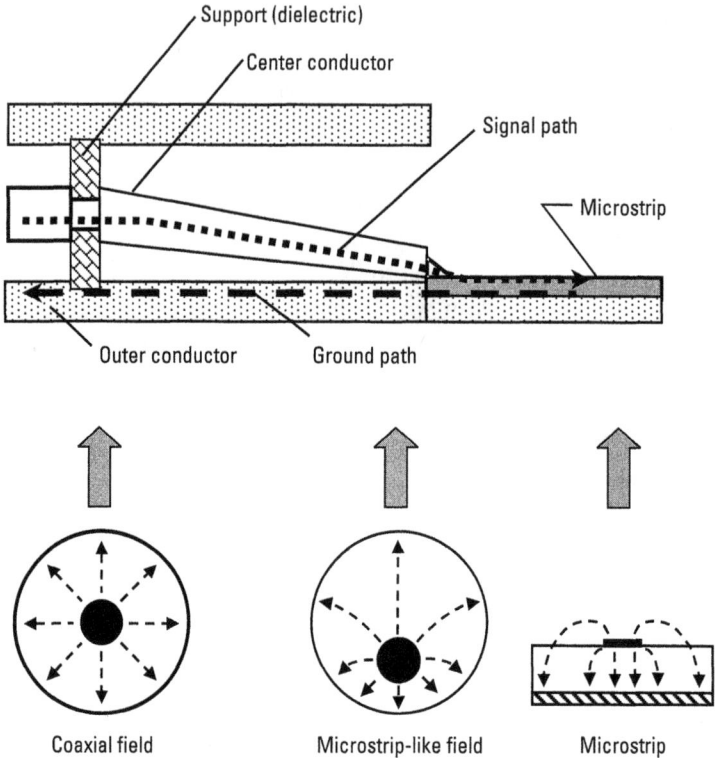

Figure 4.3 Eisenhart's microstrip-to-coax transition and electric field at different cross-sections. Signal and ground current paths are continuous and nearly equal in length. (*After:* [1].)

essentially is straight and continuous, the ground current path is more circuitous. In particular, the separation between the ground pins determines the separation of the ground and signal current paths and the upper frequency of operation for this transition. Figure 4.5 plots the input match and insertion loss versus frequency for three different ground pin spacings. Up to approximately 2 GHz, the performance of all three transitions is excellent and nearly the same. Beyond 2 GHz, the insertion loss starts to increase, with the slope versus frequency being greatest for the widest spacing. Although the degraded input match contributes to this increased loss, the primary cause is radiation, as the appearance of a pronounced resonant dip for a pin spacing of roughly two-fifths of a wavelength indicates. We recommend the ground pin spacing not exceed one-fifth of a wavelength at the maximum frequency of operation, which corresponds to an insertion loss of about 1 dB (see Figure 4.5). Consequently, for the SMA coaxial line, which has a dielectric outer diameter of about 0.180 inch (0.46 cm), the narrowest possible pin spacing limits the maximum frequency of operation to 12 GHz.

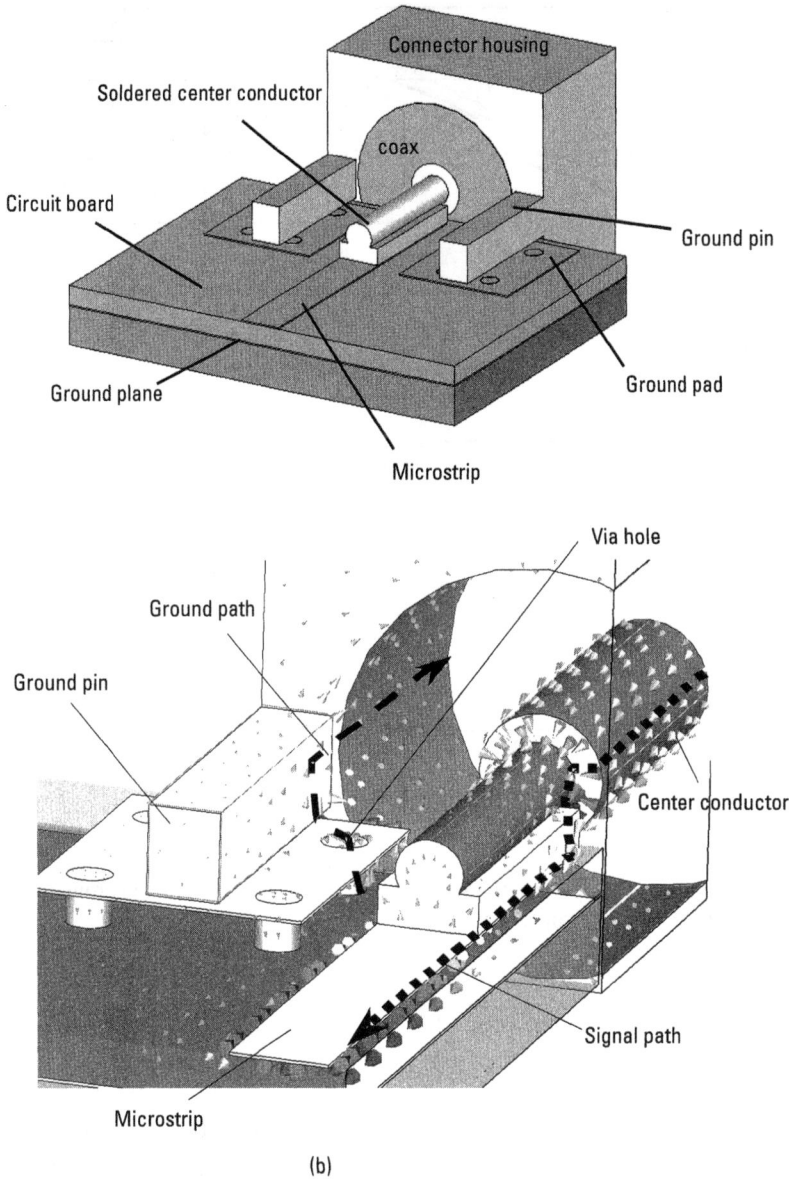

Figure 4.4 (a) Two-layer PCB mountable edge launch microstrip-to-coax transition. (b) Surface current flow reveals signal and ground current paths.

In the next example, we use the same connector in a similar transition to show how a break in the ground path can be disastrous. We also illustrate how improper circuit modeling can hide grounding problems. The circuit board in Figure 4.6 has two layers with the top layer being just 0.008 inch (0.20 mm) thick and having a relative permittivity of 3.5. A ground plane separates this

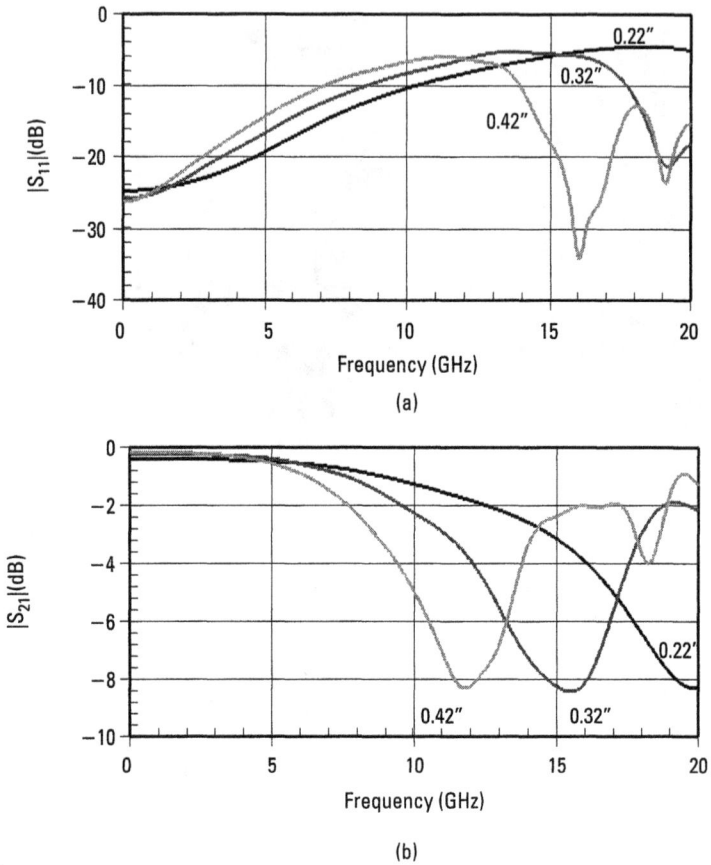

Figure 4.5 Simulated (a) input match $|S_{11}|$ and (b) insertion loss $|S_{21}|$ versus ground pin spacing for the transition of Figure 4.4.

layer from the bottom layer, which has a thickness of 0.052 inch (1.32 mm). A 50-ohm microstrip line constructed on the upper layer has a width of 0.017 inch (0.43 mm), but the coaxial center conductor diameter is 0.020 inch (0.51 mm). A good solder connection would require at least a 0.030-inch (0.76 mm) wide strip, resulting in an impedance mismatch. If we remove the ground plane under the microstrip, it will behave more like CPW, and we can widen the center trace to 0.110 inch (2.8 mm) with a comfortable 0.020-inch (0.051 mm) gap between it and the ground conductors. Thus, this transition from coaxial line to microstrip actually comprises two transitions, one from coaxial line to CPW, and the second from CPW to microstrip. As Figure 4.6(b) shows, the signal current path follows the center conductor of the coaxial line to the center conductor of the CPW, which becomes microstrip after an abrupt change in width. Although the impedances of the CPW and microstrip are the same on either side of the step in width, the end of the wide CPW center conductor is

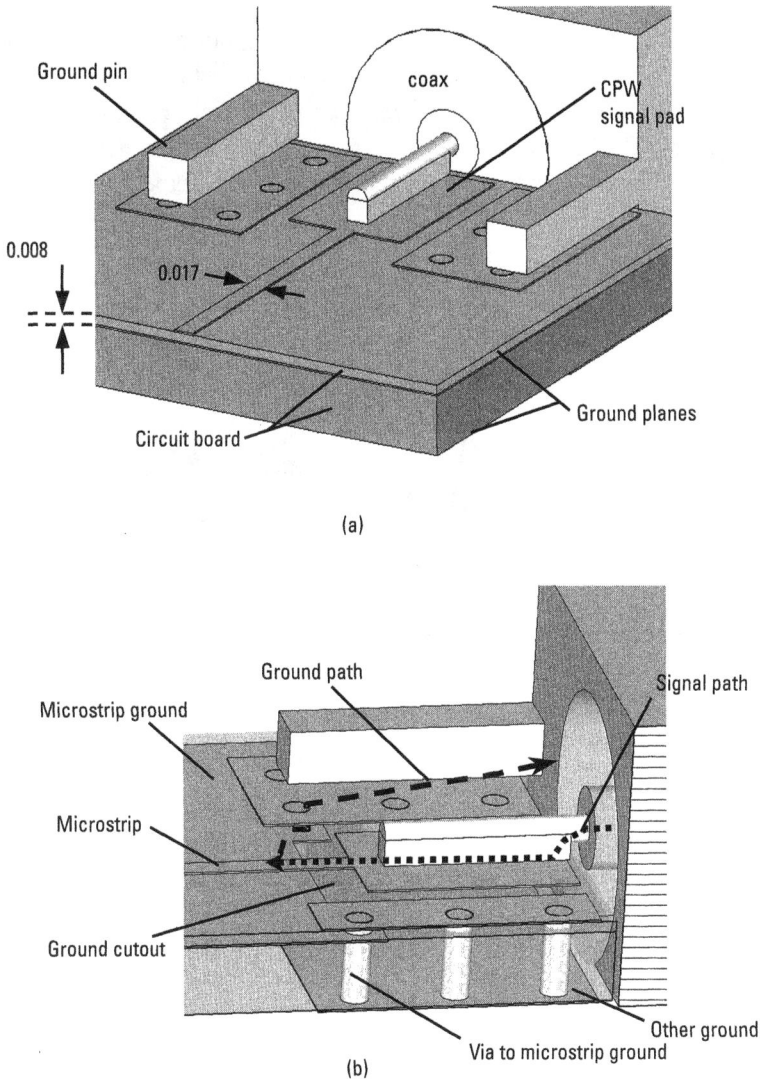

Figure 4.6 (a) Multilayer PCB mountable edge launch microstrip-to-coax transition. (b) Internal metallization and current paths. (Dimensions in inches.)

coupled capacitively to the microstrip ground plane. With the aid of an electromagnetic simulator, we can reduce the capacitance by removing a portion of the ground plane [see cutout in Figure 4.6(b)]. The increased spacing between the end of the CPW and the microstrip ground plane reduces the parasitic capacitance and extends the frequency band of the transition. The path of the ground current follows the connector ground pins onto the CPW ground pads. The vias at the end of the pads (nearest to the microstrip line) take the current down to the microstrip ground plane.

We already know that the separation of the ground pins determines the transition's maximum frequency of operation—provided the signal and ground paths are continuous. The ground current path is broken if we remove the vias interconnecting the CPW and microstrip ground planes as illustrated in Figure 4.7. At 0 Hz, with no coupling across the gap, we expect an open circuit.

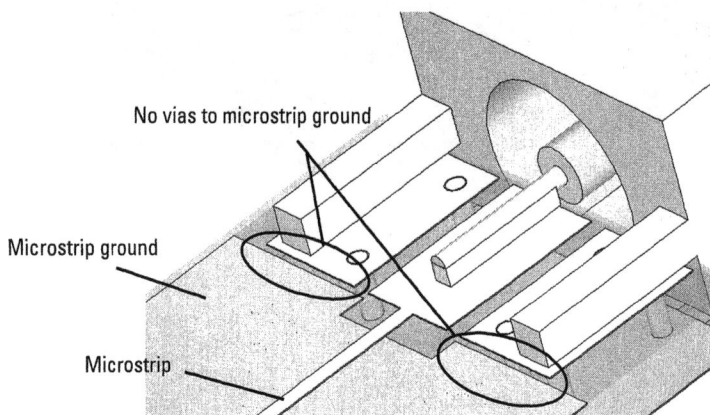

(a)

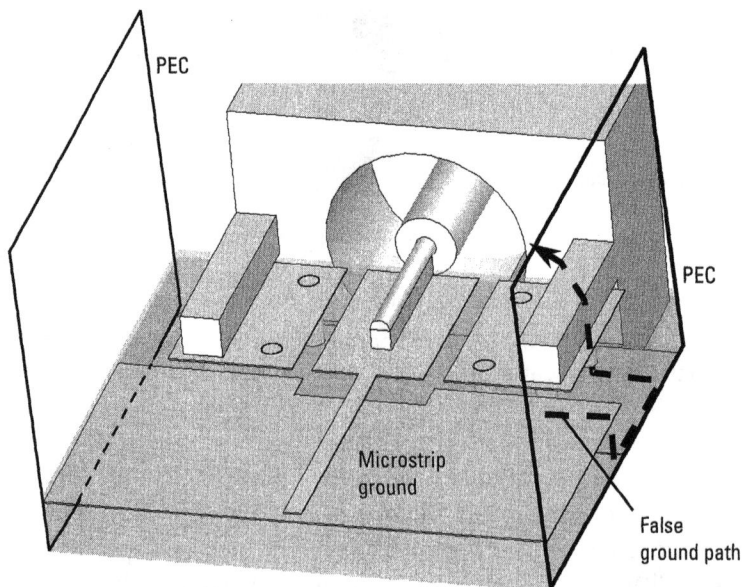

(b)

Figure 4.7 (a) Edge-launch microstrip-to-coax transition missing vias to microstrip ground. (b) A perfect electric conductor (PEC) boundary condition creates a false path for the ground current.

Figure 4.8 plots the return loss and insertion loss of the transition with and without the ground vias in place. With the vias, the return loss exceeds 20 dB and the insertion loss is less than 0.25 dB to greater than 5 GHz. Although the match still exceeds 15 dB at 6 GHz, the insertion loss has risen past 1 dB, and the knee in the insertion loss indicates the transition radiates above 5 GHz. Without the vias, the DC return loss is near 0 dB, as expected. As the capacitive impedance $\left(1/j\omega C_{gap} \right)$ of the gap decreases with increasing frequency, the ground current couples more strongly across it, and the return loss increases.

We predicted the performance for the last two examples, like many others in this book, using numerical electromagnetic analysis software. Such software analyzes a structure by subdividing it into many discrete volume elements, typically cubes or tetrahedrons. The computer solves Maxwell's equations by solving a matrix of unknowns whose size depends on the number of volume elements within the structure. We must place boundaries around the structure to limit

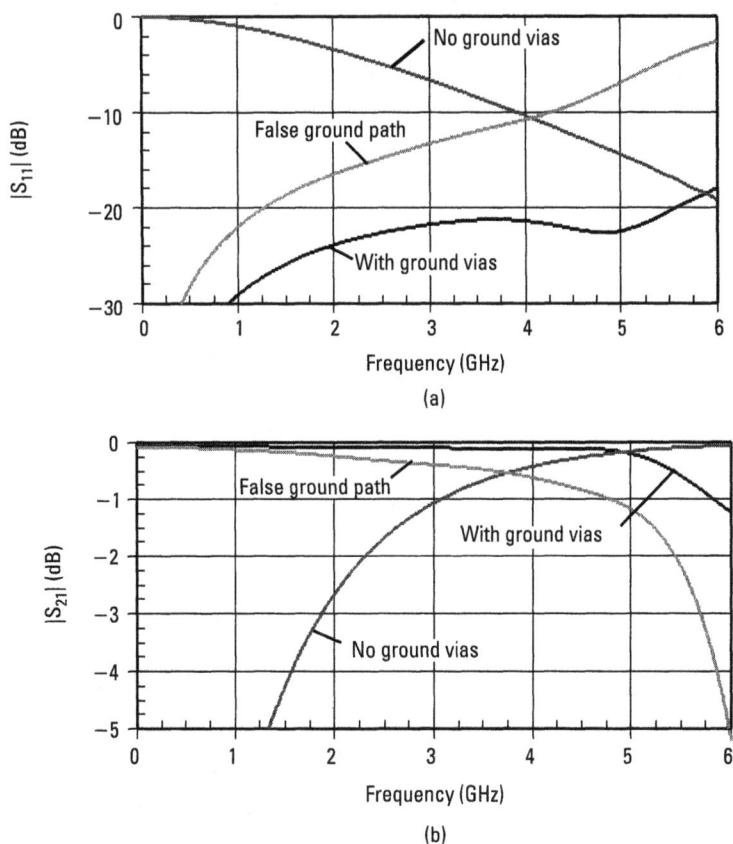

Figure 4.8 Simulated (a) input match $|S_{11}|$ and (b) insertion loss $|S_{21}|$ for the transition of Figures 4.6 and 4.7.

the number of volume elements and the solve time. In so doing, we need to be cautious. Figure 4.7(b) shows two such boundaries placed flush against the last example's transition with the missing ground vias. These PEC boundaries are too close to the structure. Because they touch the edges of both the CPW and the microstrip ground planes, they create a false ground path that bypasses the real path, which is broken. When we analyze the transition, we get an erroneous result as revealed by the data in Figure 4.8. The transition seems to be well matched with an insertion loss of only a few tenths of a decibel up to 3 GHz. Because the false ground path is longer than that in the optimized design [compare Figure 4.7(b) with Figure 4.6(b)], the upper frequency of operation is lower, but the results show no sign that the ground path is broken.

Edge-launch connectors tend to be limited in bandwidth by the separation between the signal and ground current paths. To reduce the separation, we may have to use a *connector-less transition* like that in Figure 4.9, in which the coaxial cable is soldered directly to a metal pad on the circuit board. Vias beneath the pad provide a short path directly to the microstrip ground plane. The return loss of this transition is 15 dB up to 12 GHz for a 0.047-inch (1.19 mm) coaxial cable attached to microstrip on a 0.008-inch (0.20 mm) thick Rogers 4003 substrate. Radiation is minimal at 12 GHz, as the insertion loss is about 0.25 dB.

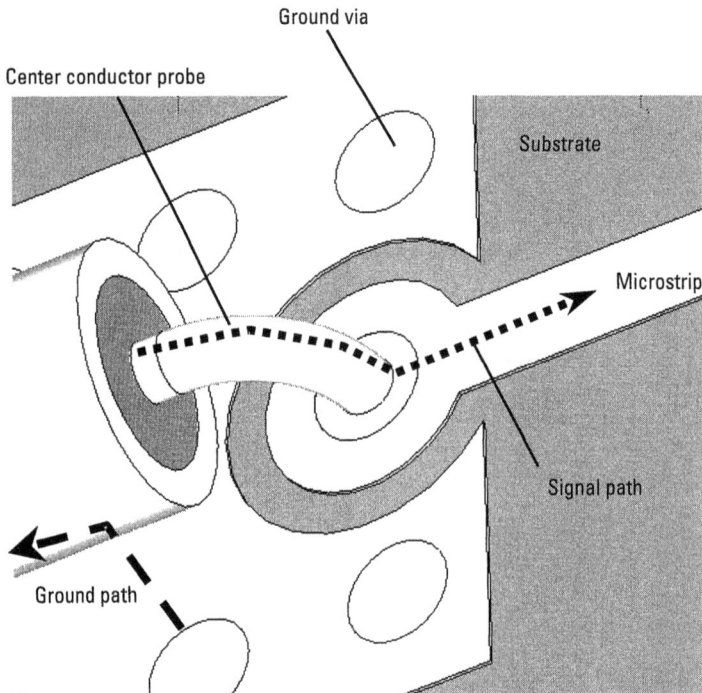

Figure 4.9 In-line, connector-less, microstrip-to-coax transition. (*After:* [2].)

4.2.2 Vertical Mounted Transitions

Edge-mounted transitions can only be placed along the edge of a circuit board. If we need a transition in the middle of a circuit board, then we can use a *vertical mounted* transition like the one in Figure 4.10(a). The vertical mounted connector comes with an extended center conductor probe and, like the edge-mounted transition, four ground pins. We insert these conductors through vias in the circuit board and solder them to metal pads on the top layer. Within the circuit board portion of the transition, the electromagnetic wave propagates in a five-wire transmission line. To enable this wave to propagate unimpeded, we must remove metallization on all intervening conductor planes as shown in Figure 4.10(a). The ground current flows up the ground pins to the microstrip ground plane. The signal current flows through the center conductor and then

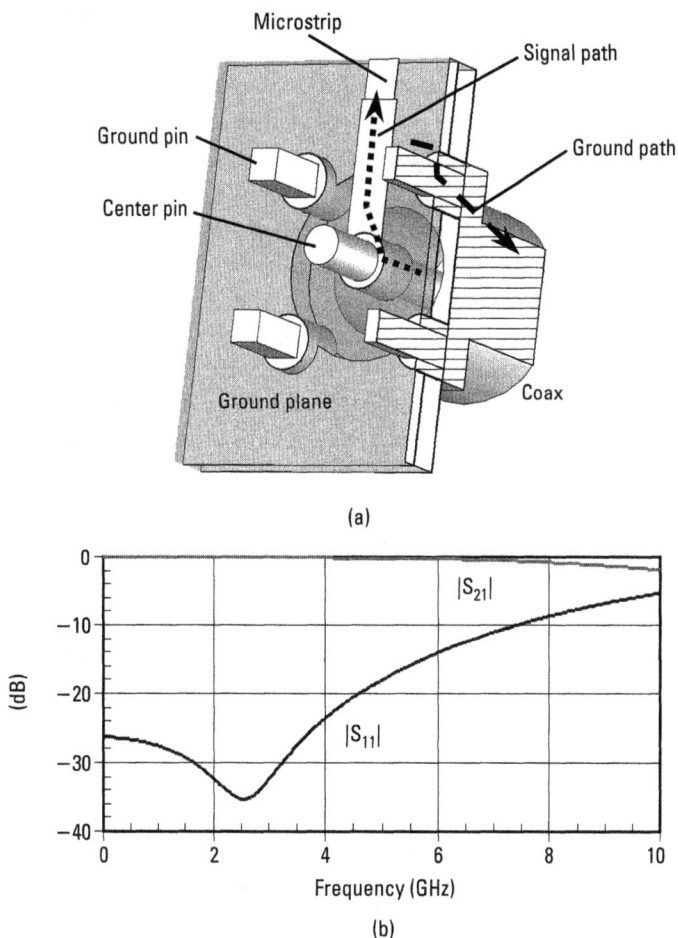

Microstrip

Signal path

Ground pin

Ground path

Center pin

Ground plane

Coax

(a)

(b)

Figure 4.10 (a) Vertical launch, SMA coax to microstrip transition, and (b) simulated performance.

turns abruptly on the microstrip layer. Because the microstrip ground plane has been cleared around the center conductor, the microstrip impedance is not well defined in the region between the center pin and the edge of the microstrip ground plane. For a ground pin spacing of 0.2 inch (0.51 cm), the return loss is greater than 15 dB to nearly 6 GHz [see $|S_{11}|$ plot in Figure 4.10(b)]. The insertion loss increases to 1 dB at 8.5 GHz. However, conservation of energy ($|S_{11}|^2$ + $|S_{21}|^2 = 1$) is satisfied up to at least 10 GHz, so mismatch rather than radiation is the cause of the increased insertion loss. As with edge-launch connectors, the ground pin to signal pin spacing establishes the maximum frequency of operation. A subminiature version P (SMP) connector, with 0.1-inch (2.5 mm) spacing between ground pins, can be made to work up to 18 GHz. An approximate rule to follow in selecting the ground pin spacing is 15% of the free-space wavelength at the highest frequency, which yields about 1-dB insertion loss.

The N-type coaxial line to microstrip transition in Figure 4.11(a) is designed to have separate low and high frequency signal current paths. Although, a type N connector is not ideal as a circuit board interface, this connector is very sturdy and can be purchased in a weatherproof version for use outdoors, as was required in this low-cost application. The figure shows the top view of the transition, including the microstrip conductors and holes for the coaxial connector ground pins. The microstrip output exits the bottom of the image. A much wider RF open circuit, a stub with vias shorting to the ground plane, points in the opposite direction. The stub directs low frequency signals such as static electricity safely to ground, protecting the circuitry on the microstrip side of the transition. We choose the stub's width and the number of ground vias to handle the power dissipated in the electrostatic discharge. The length of the stub is one-quarter of a wavelength at the center frequency of the RF operating frequency band. This open circuit will prevent any of the desired signal from propagating down the stub. However, the stub drastically reduces the transition's RF bandwidth. Without the stub, the bandwidth would extend from DC to at least the operational frequency of interest, 5.8 GHz in this case. With the stub in place, the bandwidth is very narrow, with a good match exhibited only within a few hundred megahertz bandwidth centered at 5.8 GHz. A wide stub width slightly improves the useable bandwidth and reduces the stub resistance for high levels of static discharge.

Excess insertion loss caused by radiation within a circuit board indicates that either the ground pin spacing is too great or the circuit board is too thick. The N-connector transition of Figure 4.11 has a ground pin spacing of about 0.7 inch (1.8 cm), for which the maximum frequency of operation should be 2.5 GHz according to our 15% spacing rule. To get reasonable performance at 5.8 GHz, we can bring the signal and ground currents closer together and confine the electromagnetic field better by drilling circuit board vias close to the center conductor probe as shown in Figure 4.11(b). The vias effectively replace

DC ground via

Quarter-wave
stub

Low frequency signal
Current path

RF ground
current path

Coax center pin

(a)

RF signal
current path

Substrate

Connector
ground pin

RF ground via

RF ground
current path

Microstrip

(b)

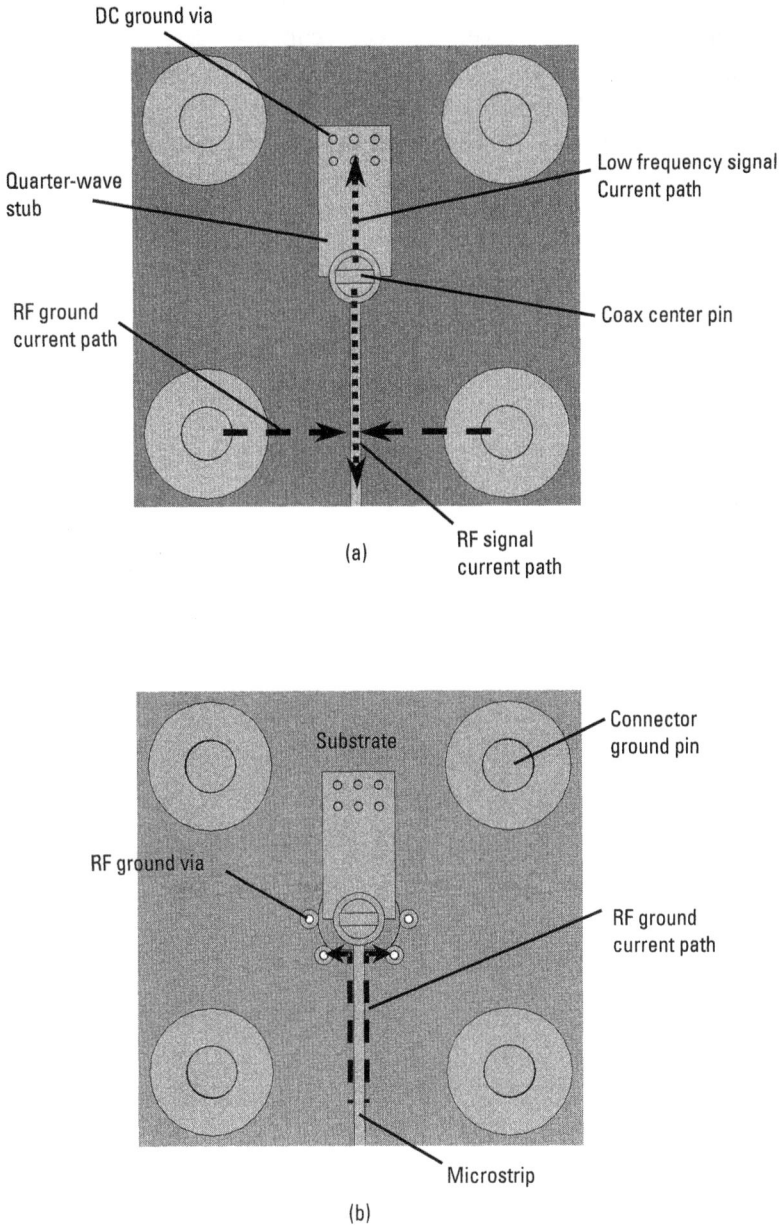

Figure 4.11 Top view of vertical launch, N-coax to microstrip transition (a) without and (b) with inner vias.

the connector ground pins as carriers of the ground current within the circuit board. Their much closer spacing increases the transition's bandwidth. For the ground path to be continuous, the vias must contact the coaxial connector outer conductor flange at the base of the circuit board.

Figure 4.12(a) shows a very low cost, connector-less coaxial line to suspended microstrip transition. This transition is used on the back of a low cost patch antenna circuit board. Suspended microstrip is required to minimize the loss in the patch antenna. The most expensive part of a coaxial transition usually is the connector, so this transition replaces the connector with four rivets, a small transition board, and a bare coaxial cable. In contrast to the connector-less

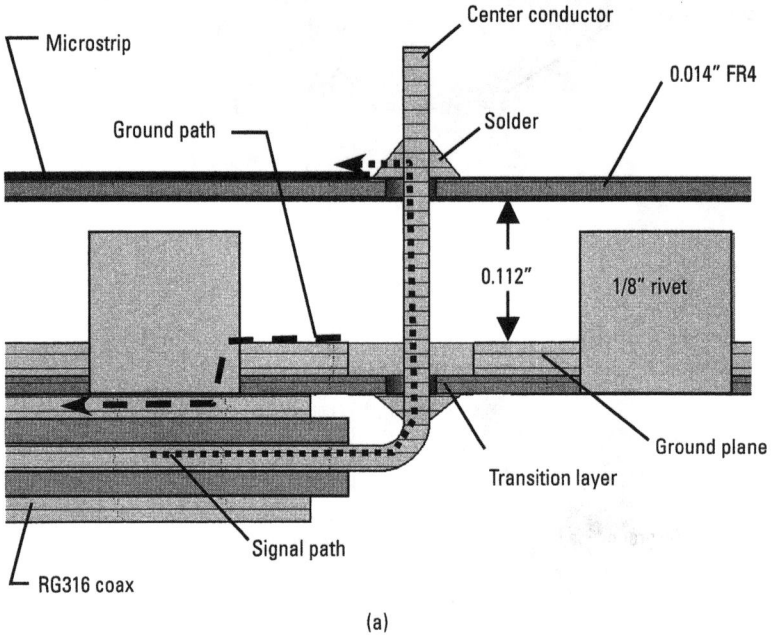

(a)

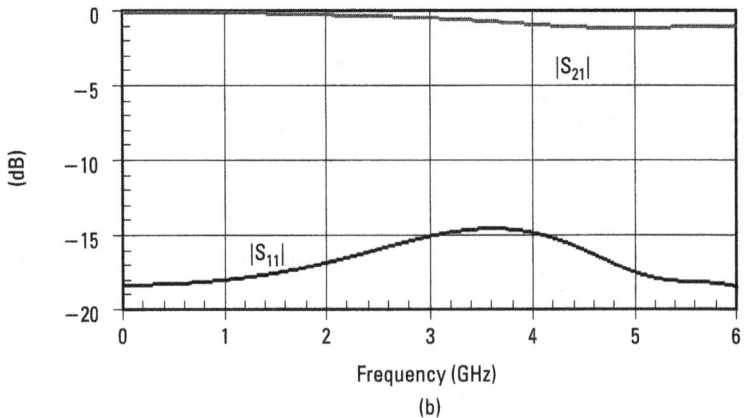

(b)

Figure 4.12 (a) Vertical launch, RG-316 coax to suspended microstrip transition, and (b) simulated performance.

transition of Figure 4.9, which uses standard microstrip, we cannot solder the coaxial cable jacket to a pad on the microstrip layer and use vias to the ground plane to complete the ground current path. Instead, we have designed this transition to be more like a vertical mounted transition, with the coaxial line center conductor protruding through a hole in the ground plane. The ground plane is a thin aluminum plate, so we cannot solder the coaxial line jacket directly to it. We use a piece of circuit board (the transition board) riveted to the ground plane as the mechanical interface. Figure 4.12(a) shows the signal current path through the center conductor to the microstrip. In the region between the microstrip ground and the microstrip, there are no ground pins or vias nearby to confine the field. Thus, for low radiation loss, we must minimize the thickness of the air suspension. For no more than 1 dB of insertion loss at the maximum frequency of operation, the air thickness plus the substrate thickness multiplied by the square root of its dielectric constant should be less than 5% of a free-space wavelength. For the transition in Figure 4.12(a), we would expect the 1-dB frequency to be about 4 GHz, as the data in Figure 4.12(b) confirms.

In summary, the separation of the signal and ground current paths is a primary limitation on the bandwidth of most coaxial to microstrip transitions. For edge and vertical mount connectors, the signal to ground pin spacing should be below one-quarter of a wavelength. In the case of transitions to suspended microstrip, the thickness of the suspension plus that of the substrate determines the maximum frequency of operation.

4.3 Waveguide to Microstrip Transitions

While coaxial line generally is used at X-band (8 to 12 GHz) and below, waveguide is used mostly above 20 GHz, where its lower loss becomes an advantage. At higher frequencies, *waveguide to microstrip transitions* replace transitions to coaxial line: they serve as interconnects between sealed modules or between modules and antennas. These transitions can be designed to operate at very high frequencies, in excess of 100 GHz. We know from Chapter 3 that waveguide, being formed from a single conductor, propagates a dominant mode, usually of the TE configuration, that has a cutoff frequency below which the waveguide is highly attenuative to EM signals. Most transitions are designed to operate within the frequency band of dominant mode propagation only, which is at most 2:1 for rectangular waveguide and 1.3:1 for circular waveguide.

As compared with coaxial line, waveguide modes have impedance characteristics that tend to make transition design more challenging. The *dispersion* (nonlinear relationship between propagation constant and frequency) of waveguide means that the impedance of each of its modes changes with frequency. In addition, the impedances of standard waveguides are much greater than 50 ohms, typically a few hundred ohms for TE modes. Consequently, the bandwidth for most waveguide to microstrip transitions rarely reaches the full dominant mode bandwidth.

4.3.1 Orthogonal Transitions

Like coaxial to microstrip transitions, waveguide to microstrip transitions can be edge or orthogonally configured, with the latter being more compact. Figure 4.13 shows an *orthogonal* or *right angle transition* to circular waveguide. The microstrip protrudes through a narrow channel into the waveguide. The channel, a section of waveguide itself, has sufficiently small width and height so that only the microstrip mode can propagate within the band of operation. All waveguide modes are cutoff. The microstrip ground plane ends at the edge of the main waveguide, thus enabling the signal current on the microstrip probe (see Figure 4.13) to radiate inside the waveguide as an antenna. Since the probe radiates equally well up and down the waveguide, we block the undesired direction with a waveguide short circuit or *backshort,* situated approximately one-quarter of a wavelength away from the probe. In effect, the backshort creates an RF open circuit at the plane of the probe so that the backward wave radiated by the probe adds in phase to the forward wave exiting the transition.

For this type of transition to operate well, the microstrip ground plane must make electrical contact with the channel floor. The currents that flow on

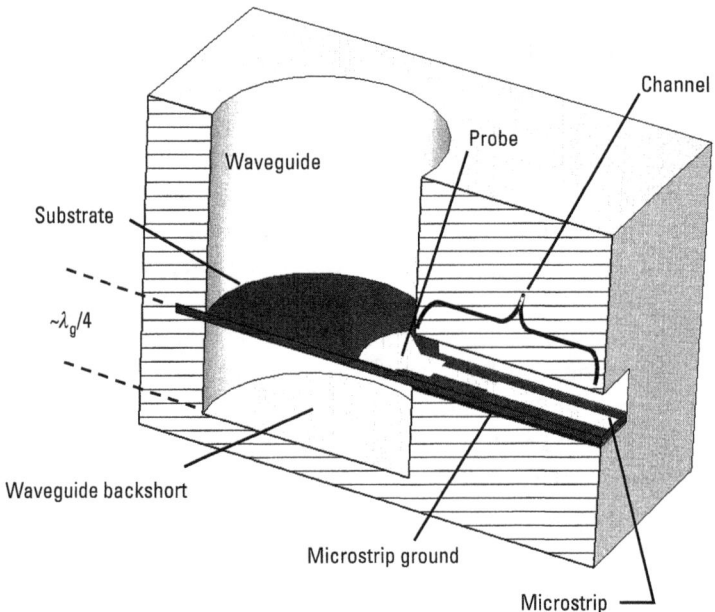

Figure 4.13 Orthogonal microstrip to circular waveguide transition.

the microstrip ground plane must be able to flow uninterrupted into the wave-guide conductor. Figure 4.14 helps to explain why even a minute gap between the microstrip ground plane and the channel surface can seriously degrade the transition's performance. The figure shows a cross-sectional view of the channel containing the microstrip. There is a gap of just one-thousandth of an inch (0.025 mm) between the microstrip ground plane and the channel floor. Figure 4.14(a) shows the electric field distribution of the usual quasi-TEM mode. If there were no gap, this mode would be the only one that could propagate in the channel. However, the gap separates the microstrip ground plane and the channel surface, so that a parallel-plate waveguide mode can propagate also as shown in Figure 4.14(b).

To understand the effect of the gap, we designed the transition shown in Figure 4.13 to operate from 54 to 62 GHz and then analyzed it with and without the gap present. The results are plotted in Figure 4.15. With no gap, the return loss is better than 25 dB over the entire frequency band and the insertion loss is nearly zero. With a 0.001-inch (0.025 mm) gap underneath the substrate ground plane, the return loss degrades at least 10 dB, and the insertion loss increases to 1 dB. As Figure 4.15(b) shows, this lost energy is converted to the parallel-plate mode in the gap and then reflected at the main waveguide's interface with the channel back towards the microstrip input. Clearly, a physical connection between the waveguide and microstrip ground must be made for this transition to perform well. The connection can be made with conductive adhesives such as silver epoxy, but permanent contact is not necessary. Figure 4.16 shows a photograph of a microstrip to waveguide transition that has been

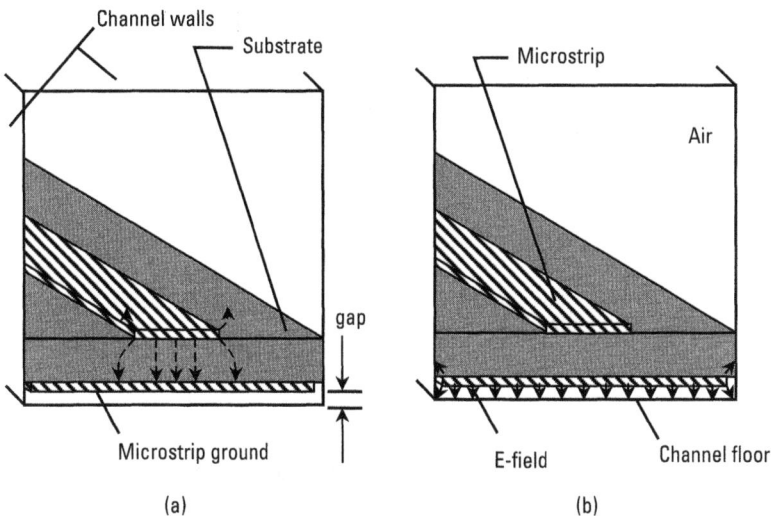

Figure 4.14 Electric field modes in microstrip port with a 0.001-inch (0.025 mm) gap beneath the substrate: (a) quasi-TEM mode; and (b) parallel-plate waveguide mode.

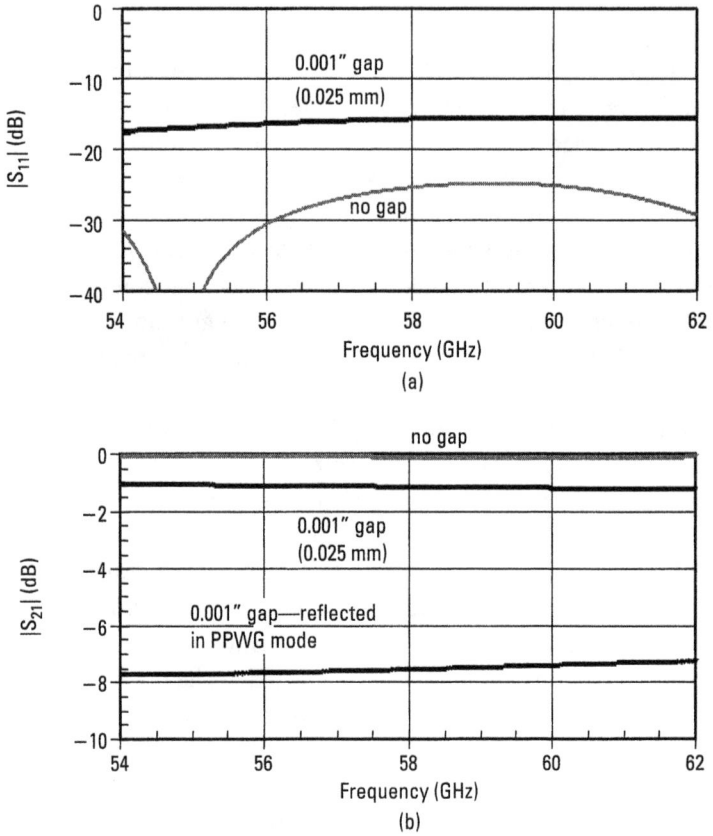

Figure 4.15 Microstrip to circular waveguide transition. Simulated (a) return loss and (b) insertion loss with and without a gap under the substrate.

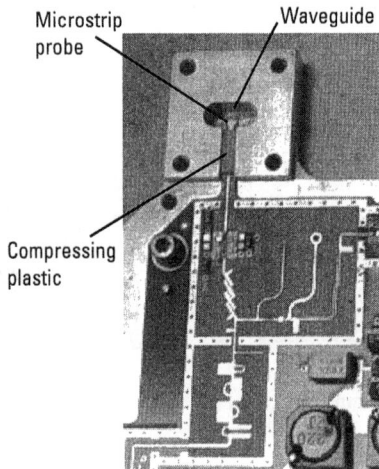

Figure 4.16 Microstrip to rectangular waveguide transition integrated with a transceiver circuit board.

integrated into a communications transceiver circuit board housing. A metal cover containing the waveguide backshort, the channel sidewalls, and top surface fastens on top of the circuit board. We need to be able to remove the circuit board for testing and rework, so we enforce contact of the probe to the floor of the channel by jamming a piece of plastic or other bendable dielectric material between the top of the microstrip substrate and the top of the channel.

The microstrip circuit in the previous example was fabricated on a single layer of dielectric so that the microstrip ground plane could mount directly to the channel surface. The circuit in the transition of Figure 4.17 is fabricated on a

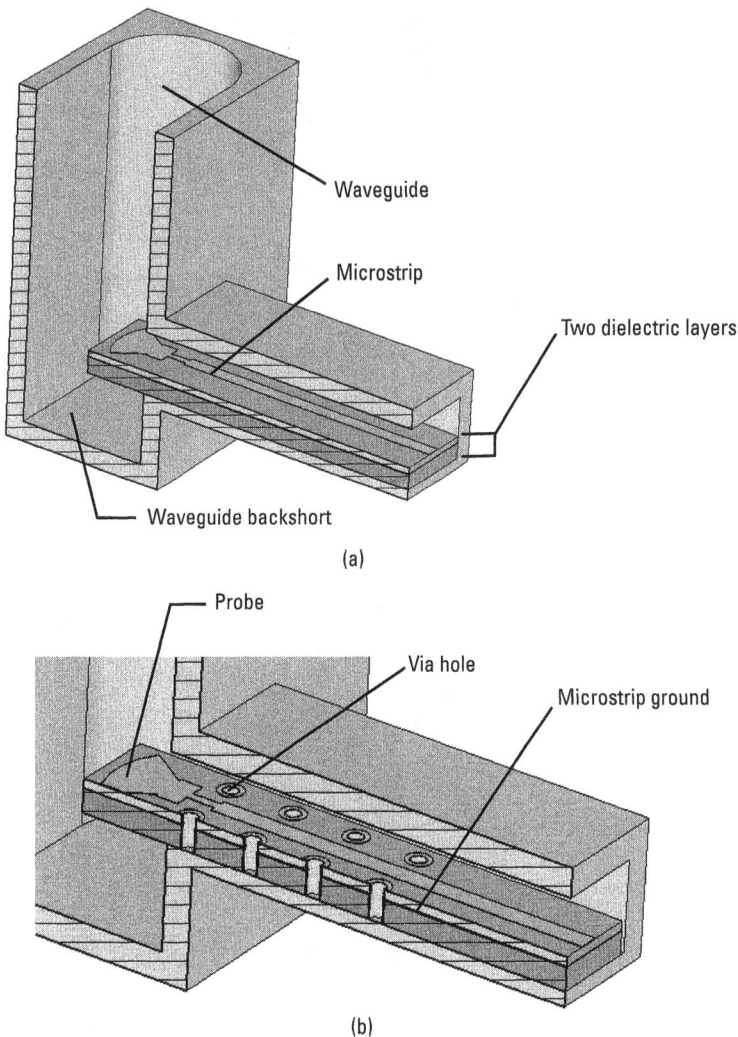

(a)

(b)

Figure 4.17 (a) Transition between microstrip on two-layer PCB and full-radius rectangular waveguide. (b) PCB with mode suppression grounding vias.

two-layer substrate, and the microstrip ground plane is located between the two layers. Such a transition will be lossy unless we find a way to insure good electrical contact between the microstrip ground plane and the channel floor. Figure 4.18 shows the microstrip port of the transition. In this example, the microstrip ground plane touches the vertical walls of the channel to form a small rectangular waveguide enclosing the lower dielectric of the circuit board. Besides the quasi-TEM microstrip mode in Figure 4.18(a), a waveguide TE_{10} mode can propagate in the lower dielectric, as shown in Figure 4.18(b). This mode would

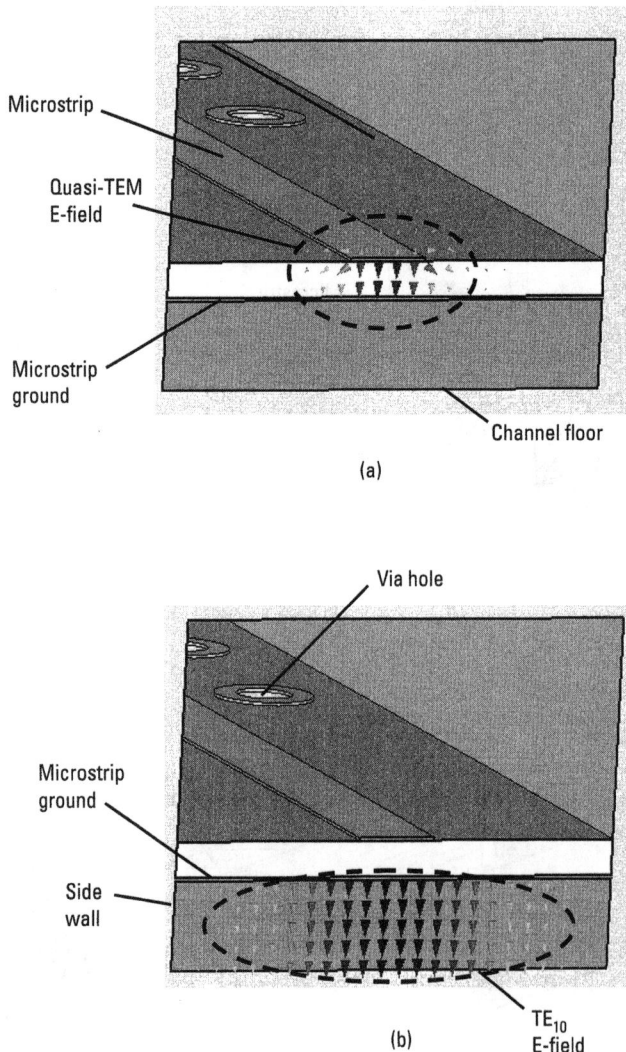

Microstrip

Quasi-TEM
E-field

Microstrip
ground

Channel floor

(a)

Via hole

Microstrip
ground

Side
wall

TE_{10}
E-field

(b)

Figure 4.18 Microstrip port of transition showing electric field of (a) desired quasi-TEM mode and (b) unwanted TE_{10} mode.

be in cutoff were it not for the dielectric, which reduces the cutoff frequency by the square root of its dielectric constant (typically 3 to 4).

A simple way to form an uninterrupted ground current path between the microstrip ground plane and the channel floor is to drill via holes through the circuit board as shown in Figure 4.17(b). Several pairs of these via holes will reject the TE_{10} mode. To demonstrate, we optimized the transition's dimensions for operation over the 24- to 28-GHz band, as shown by the plots in Figure 4.19. With four pairs of grounding vias, the return loss exceeds 20 dB and the insertion loss is insignificant. On the other hand, without these vias the return loss decreases to 10 dB, and the insertion loss increases to 1.5 to 2 dB just above the TE_{10} mode cutoff frequency (25.8 GHz). Most of this loss occurs as conversion from the incident microstrip quasi-TEM mode to the reflected TE_{10} mode [see Figure 4.19(b)]. Grounding vias are an effective solution, but only if they are made to touch the channel floor using one of the methods we described previously for the single-layer circuit board.

4.3.2 End-Launched Transitions

Waveguide-to-microstrip transitions can be end-launched also. Figure 4.20 shows an end-launched microstrip to parallel-plate waveguide transition.

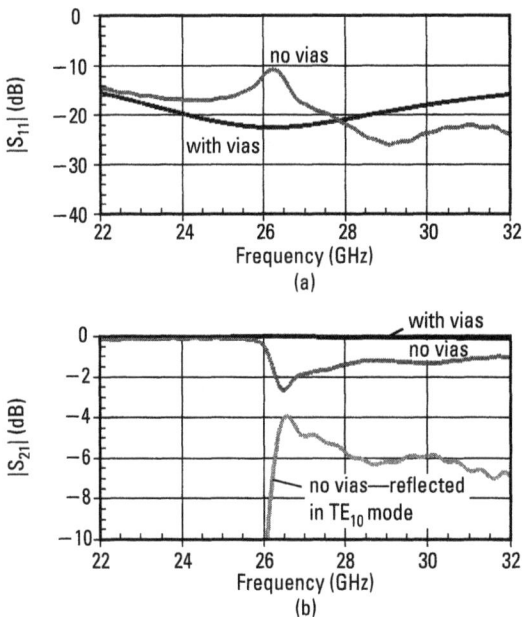

Figure 4.19 Microstrip on two-layer PCB to rectangular waveguide transition. Simulated (a) return loss and (b) insertion loss with and without mode suppression vias through substrate.

Because both transmission lines propagate the TEM mode, the bandwidth of this transition is very broad and can exceed two octaves [3]. Since parallel-plate waveguide has separate signal and ground conductors, each current component must flow into a separate waveguide wall. Both the microstrip and its ground plane must make physical contact with the waveguide. For a transition like that in Figure 4.20, such contact is essential for operation to 0 Hz since DC current cannot flow across a gap without arcing. This particular transition is also termed a *balun* (i.e., balanced-unbalanced), a structure that transforms an unbalanced transmission line having one conductor at ground potential (the ground plane) and the other referenced to it, to a balanced transmission line, for which the two conductors are at equal but oppositely polarized potentials. In Figure 4.20, the microstrip is the unbalanced transmission line, and the parallel-plate waveguide is balanced. The transformation from an unbalanced to a balanced structure

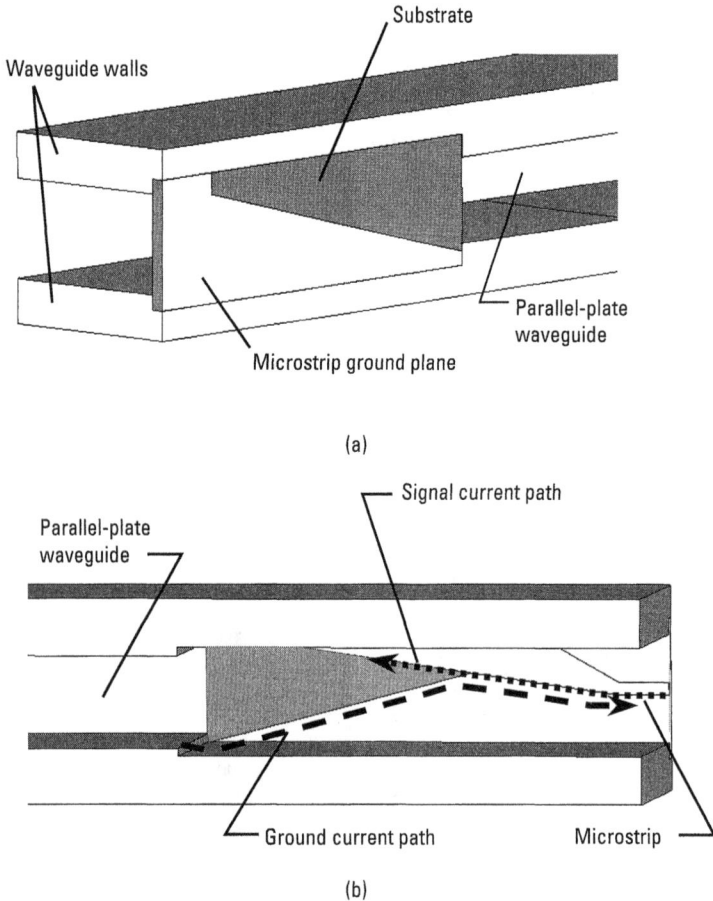

(a)

(b)

Figure 4.20 Microstrip to parallel-plate waveguide transition: (a) microstrip ground-plane side; and (b) microstrip side with ground plane showing through substrate.

occurs on the microstrip substrate as the flaring signal and ground plane conductors force the signal and ground currents to diverge [see Figure 4.20(b)]. This transition is unusual in that it requires the currents to separate and not flow parallel to each other.

4.4 Microstrip Transitions to Other Planar Transmission Lines

While microstrip transitions to coaxial line and waveguide enable electromagnetic signals to travel from one microwave component to another, transitions often are required within a single circuit board. For example, the coaxial line to microstrip transition in Figure 4.6 actually is a transition from coaxial line to CPW. A second transition like that shown in Figure 4.21 transforms from CPW to microstrip. CPW and microstrip both propagate the quasi-TEM mode, so to design this transition we first match the characteristic impedances of the two transmission lines. The signal current can flow almost unimpeded from the microstrip line to the center conductor of the CPW. However, the microstrip and CPW ground planes are on different layers. The CPW electric field is polarized or oriented parallel to the substrate surface, which is perpendicular to the plane of the microstrip field. Thus, the electric field must rotate 90° as it passes through the transition, and it is the path ground current takes through the transition that causes the field rotation. Starting from the microstrip line, the ground current flows under the microstrip until it reaches the transition. It then flows perpendicularly outwards towards the via holes that bring it up to the CPW ground plane; at the same time, the electric field lines, which terminate on the ground current, gradually rotate. We want to minimize the distance over which the signal and ground currents flow perpendicular to each other to minimize radiation, which means we should locate the vias in the CPW ground as near as possible to the center conductor.

Another transition from microstrip to CPW is shown in Figure 4.22 [4]. In this example, it is the signal current that changes conductor layers. It flows to the end of the microstrip line and down a via hole to the microstrip ground plane. At the bottom of the via, the current splits in half, with each half flowing along a slot line. At the same time, the microstrip ground current flows until it reaches the slot, then splits, with each half flowing along the opposite side of the slot line. The slot lines unite at the top of Figure 4.22 to form the CPW. In contrast to the transition of Figure 4.21, the signal and ground currents stay in close proximity and flow in parallel directions nearly always, the only exception occurring when the signal current flows through the via. However, this transition involves more convoluted paths for both currents to follow and includes a transition to slot line as an intermediary step. So, although no one major discontinuity exists to limit the transition's performance, there are a number of small

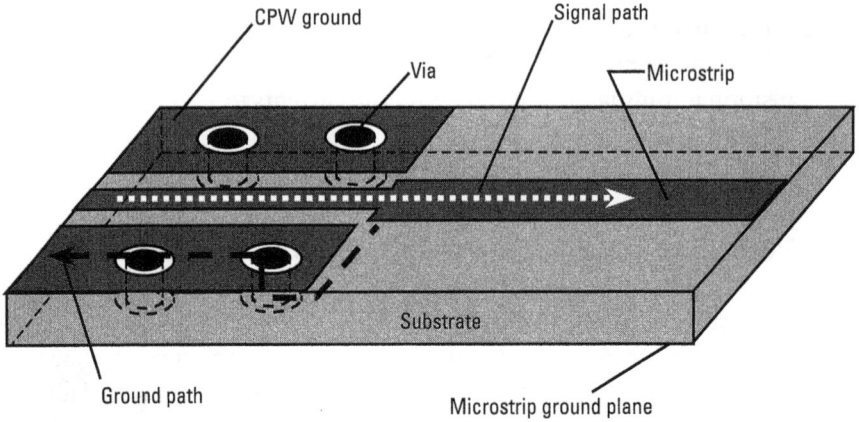

Figure 4.21 Microstrip to coplanar waveguide transition with ground current changing layers.

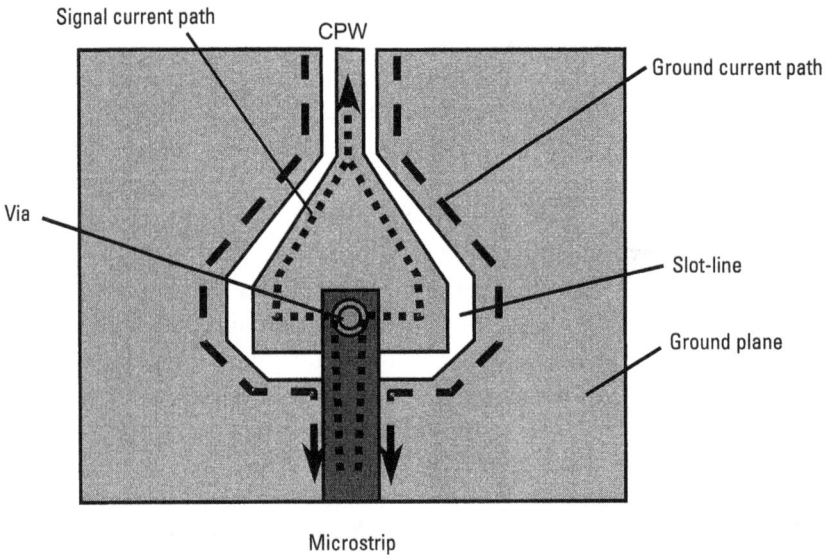

Figure 4.22 Microstrip to CPW waveguide transition signal current changing layers. (*After:* [4].)

ones. This transition will work with or without the via in the signal path. With the via in place, the bandwidth will extend down to 0 Hz. Without it, DC operation is not possible since coupling between the microstrip and ground plane currents is required, but significant bandwidth at millimeter-wave wavelengths has been demonstrated [4].

Circuit boards with many layers sometimes require vertical transitions that bring RF signals from one layer down or up to another RF layer. Such a transition from microstrip to microstrip through a three-layer circuit board is pictured in Figure 4.23. A microstrip and its ground plane are formed on the upper and lower dielectric layers, and a middle dielectric layer separates them. The transition is similar to the vertical mount coax to microstrip transition we discussed in Section 4.2.2. A multiwire transmission line formed by three vias provides the signal and ground current paths through the circuit board. The spacing and diameters of the vias are chosen to match the 50-ohm impedance of the microstrip lines. As with the coaxial transition, we must remove metal from the ground planes so as not to short circuit the TEM field, as the currents flow through the board. The signal current has to travel through all three circuit board dielectric layers, but its path is direct. Because the ground plane has been cleared under the strip to make way for the TEM field, the ground current follows a more lengthy path (it follows a similar path in the microstrip to CPW transition of Figure 4.21). In particular, the ground current and signal current paths are not parallel in the region where the ground current follows the circumference of the cleared area in the microstrip ground plane. We can shorten the ground current path if we reduce the diameter of the clearing, but then we will have to reduce the spacing of the vias and change the three-wire line impedance. To maintain the same line impedance, (3.14) requires the via diameter to be reduced also. We have designed this transition on a circuit board made from three layers of FR4, 0.062 inch (1.57 mm) thick, using vias that are 0.013 inch

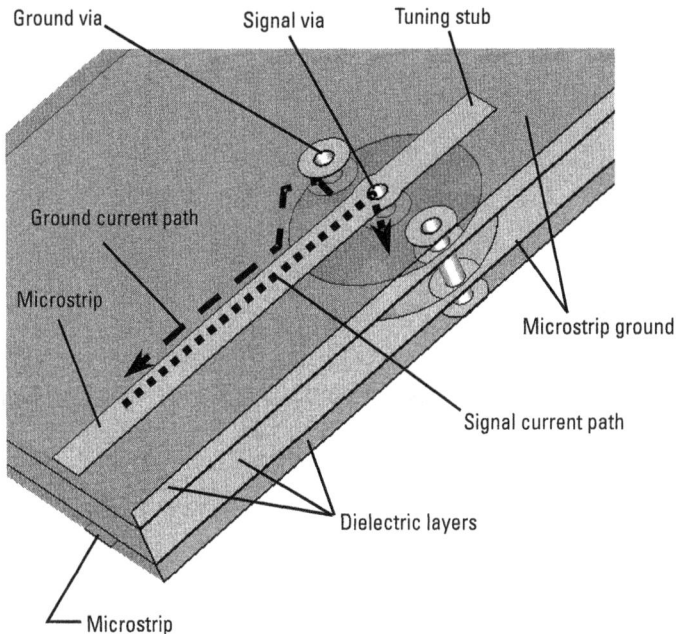

Figure 4.23 Microstrip to microstrip transition through a three-layer circuit board.

(0.33 mm) in diameter with a center-to-center separation of 0.052 inch (1.32 mm). The radius of the cleared area on each ground plane is 0.060 inch (1.52 mm). The return loss exceeds 25 dB to nearly 5 GHz.

4.5   Transitions in Microwave Test Circuits

Microwave test circuits often require transitions like the ones we have discussed so far. Test circuits are used to evaluate the performance of active microwave devices such as amplifiers, mixers, and switches. With a microwave device in place, we measure the test circuit's S-parameters with a network analyzer over a frequency range of interest. Most analyzers have coaxial or waveguide ports to interface with test circuits, while most test circuits use microstrip lines as the interface to the semiconductor device chip. An edge-launch transition to coaxial line such as Eisenhart's provides the interface between the test fixture and the network analyzer.

At millimeter-wave frequencies, we sometimes use waveguide-based test fixtures such as the one shown in Figure 4.24. This is a reusable fixture, in that we mount the active device along with the substrates required to deliver its DC

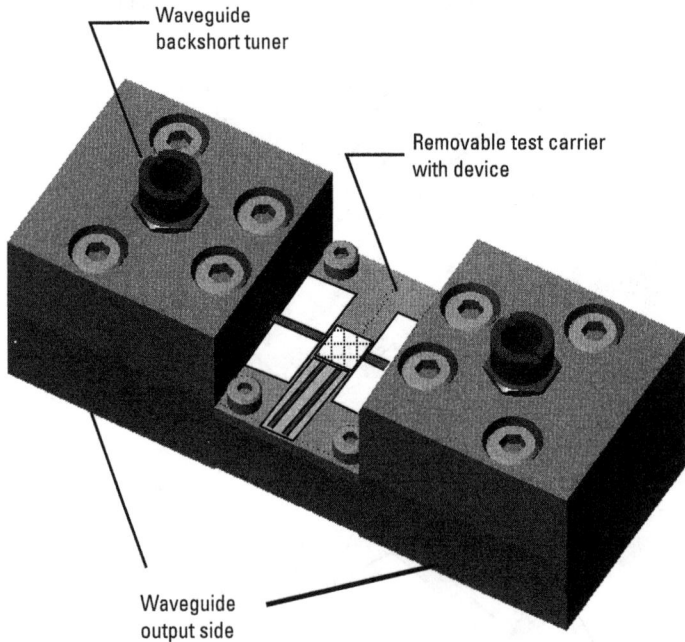

Figure 4.24 Millimeter-wave device test fixture with removable test carrier uses two adjustable waveguide to microstrip transitions to interface with measurement equipment.

bias current and RF signals on a thin metal carrier. We fasten the carrier to a metal block and place a waveguide to microstrip transition at each end of the carrier to provide the RF interface to the test equipment. Because we want to be able to remove the carrier and replace it with another, the carrier and waveguide transition require separate microstrip circuits, which are connected with wire bonds, as shown close up in Figure 4.25. The carrier and metal block are separate parts, so a small gap exists between them, which is exaggerated in the figure for clarity. There are number of ways to bridge the gap. In Figure 4.25(a), the two microstrip lines are cut flush with the edges of the metal block and carrier. A bond wire crosses the gap between the signal conductors, completing the signal current path. The ground current path is much too long, at least twice the thickness *d* of the carrier. At 60 GHz, we measured about 5 dB of insertion loss with a 0.050-inch (1.27 mm) thick carrier. Figure 4.25(b) shows a recommended way to make the connection between the microstrip lines. We eliminate the long ground current path by running one of the microstrip substrates—in this case the carrier's—across the fixture gap. Since the gap between precisely machined parts is very small, the substrate need be extended only slightly past the edge of

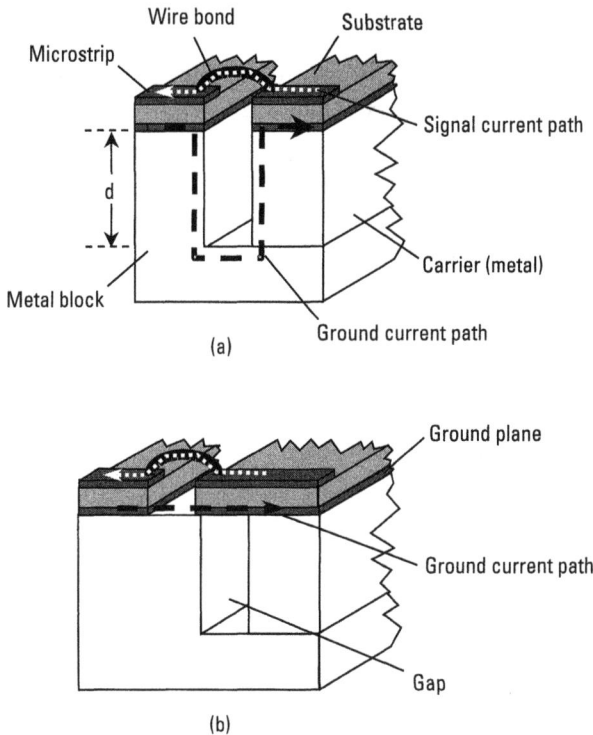

Figure 4.25 (a) Gap between circuit boards forces ground current to take a much longer path than signal current. (b) Circuit board ground plane carries ground current across gap.

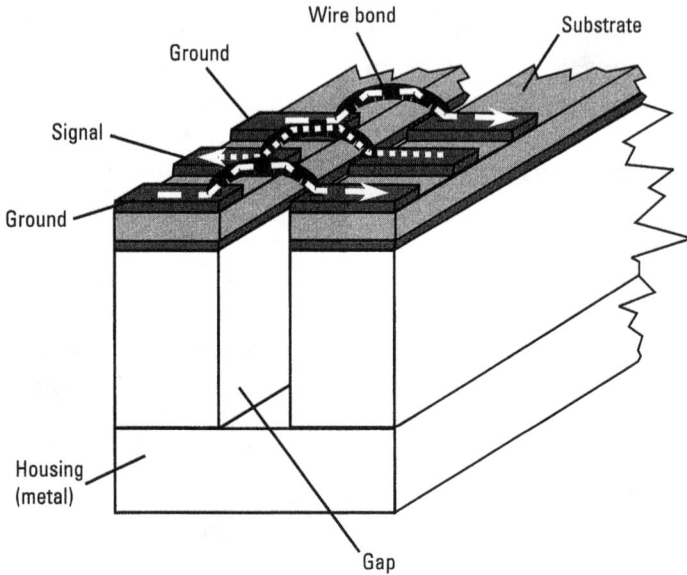

Figure 4.26 Coplanar waveguides connected by wire bonds provide paths for signal and ground currents to cross a gap between circuit boards.

the fixture to improve performance dramatically. With two wire bonds and even a relatively large gap of 0.005 inch (0.13 mm), the predicted insertion loss is 0.9 dB and the return loss is an acceptable 8 dB at 60 GHz. Our measurements indicate similar results.

Figure 4.26 shows another way to bridge a gap in a test fixture. In this case, we have transitioned to CPW and then used wire bonds to provide the signal and ground current path connections over the gap.

References

[1]   Eisenhart, R. L., "A Better Microstrip Connector," *IEEE MTT-S Int. Microwave Symp. Digest*, 1979, pp. 318–320.

[2]   Holzman, E. L., "Microstrip Transitions," in *Encyclopedia of RF and Microwave Engineering*, Vol. 4, K. Chang, (ed.), New York: Wiley, 2005.

[3]   Holzman, E. L., "A Wide Band TEM Horn Array Radiator with a Novel Microstrip Feed," *IEEE International Conf. Phased Array Systems Tech. Digest*, 2000, pp. 441–444.

[4]   Ellis, T. J., et al., "A Wideband CPW-to-Microstrip Transition for Millimeter-Wave Packaging," *IEEE MTT-S International Microwave Symp. Digest*, Vol. 2, 1999, pp. 629–632.

5

Active Microwave Devices and Circuits

Active microwave devices are the workhorses of microwave systems. They perform a variety of key functions, including the transformation of DC current to RF current (oscillation), the conversion of a signal at one frequency to another (multiplication, mixing), the redirection of a signal (switching), and the intensification and replication of a signal (amplification). Of the large variety of active microwave devices in use, we focus on the two most common, the solid-state diode and field effect transistor. We review the operational characteristics of these two devices and then explain the importance of grounding in circuits where they are used. With the aid of a number of examples, we show that the performance of a circuit can be optimized when the active device's ground path impedance is minimized and modeled accurately. We also describe techniques for grounding active devices. The chapter concludes with a discussion of grounding for multidevice modules and circuit boards.

5.1 Introduction

The previous two chapters focused on transmission lines and related *passive circuits* whose predominant function is to transport RF and microwave signals from one point to another with as little loss as possible. The sources of the signals in such circuits are *active* microwave devices. The transfer characteristic of an active device is a function of its DC operating point, its frequency band, and the power level of its input signal. An active microwave device usually is embedded in a passive matching circuit, which optimizes its performance and provides the device's grounding path. We call the combination of an active microwave device and its passive circuit an *active microwave circuit.* A well-characterized ground path having low impedance is critical for achieving the desired performance from such a circuit.

Two popular active microwave devices are the semiconductor diode, a two-port device, and the microwave field effect transistor (FET), a three-port device. The diode's nonlinear current-voltage characteristic is useful for switching and frequency generation and conversion. The transistor, a voltage controlled current source, is most often used as an amplifier, but it also may be used for other purposes, including mixing, multiplication, and oscillation. Many books are available that describe how to design circuits with these devices. In this chapter, our goal is to explain the characteristics of diodes and transistors sufficiently to illuminate how grounding influences their behavior. Table 5.1 summarizes the effect of ground path impedance on the active circuits discussed in this chapter.

5.2 Microwave Diodes

A *diode* is a nonlinear device that conducts current with low resistance when a positive bias is applied across its terminals. When a negative bias is applied, a diode acts as a voltage-dependent capacitor. The smaller the diode's resistance and capacitance, the higher in frequency it can operate. Some common applications for diodes are as limiters, switches, detectors, reactive tuning elements, frequency multipliers, and mixers.

5.2.1 Diode Operation

There is a variety of semiconductor diode types, but fundamentally, all operate like the *pn* junction shown in Figure 5.1. As we explained in Section 2.2.3, the

Table 5.1
Effects of Ground Resistance and Inductance on Active Circuit Performance

Circuit Type	Device Type	Effect of Ground Resistance	Effect of Ground Inductance
Switch	Diode	Reduced isolation	Reduced isolation Increased insertion loss
Mixer	Diode	Not significant for typical resistances	Increased conversion loss
VCO	Diode or MESFET	Not significant for typical resistances	Tuning bandwidth shifts
Amplifier	MESFET	Reduced gain, P_{out}, increased stability	Reduced gain, increased noise figure, reduced stability

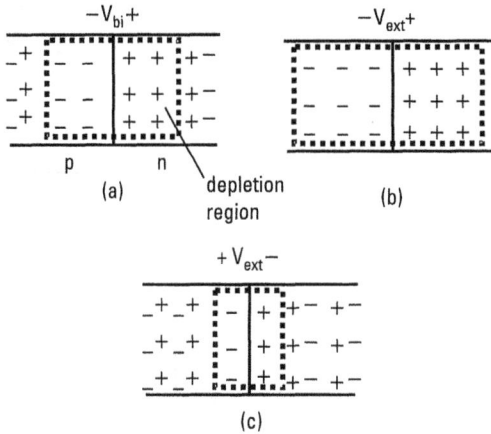

Figure 5.1 (a) *pn* junction in equilibrium has a built-in potential V_{bi} across its depletion region. (b) A reverse bias increases the width of the depletion region. (c) A forward bias reduces the width of the depletion region.

balance between the forces of diffusion and electrostatics establishes a *built-in* negative potential across the junction, which must be overcome before current can flow. The built-in potential is established within an area on either side of the junction, called the *depletion region* that is devoid of free charges [see Figure 5.1(a)]. The depletion region capacitance serves to block the flow of current through the junction. Because the junction is electrically small even at microwave frequencies, its capacitance can be calculated from electrostatic theory given the properties of the p and n materials [1].

If, as depicted in Figure 5.1(b), we subject the diode to an external *reverse bias* DC voltage that reinforces the built-in potential, then more charges diffuse across the junction, leaving an enlarged depletion region and increased junction capacitance. Only a minute amount of current, the *reverse leakage current,* normally flows across a reverse biased diode. The junction capacitance will increase with increasing reverse bias until the *reverse breakdown* condition is reached. At breakdown the current flow increases dramatically. In the discussions that follow, we assume that this condition is never reached.

The external voltage in Figure 5.1(c) *forward biases* the diode and reduces the depletion region and junction capacitance. Once the forward bias exceeds the built-in potential (less than a volt for most microwave diodes), the current flowing through the diode will increase exponentially with the voltage across its terminals.

Figure 5.2(a) shows the schematic symbol of a grounded diode. The junction voltage, $V_J = V_{ext} + V_{bi}$, is referenced to ground, with V_{bi} being less than zero, and V_{ext} being greater than zero for forward bias. The *ideal diode equation* relates the current flowing through the diode to its junction voltage

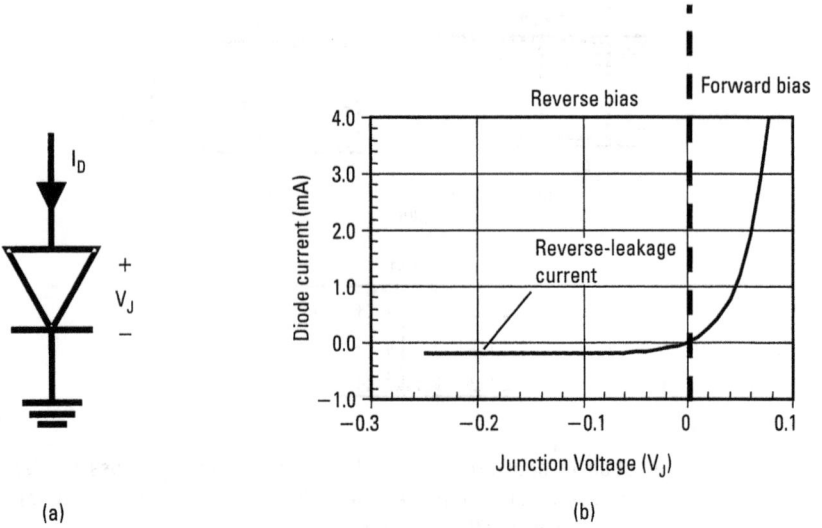

Figure 5.2 (a) Diode symbol and (b) ideal diode current versus junction voltage.

$$I = I_0\left(e^{qV_J/kT} - 1\right) \tag{5.1}$$

where I_0 is the reverse leakage current, V_J is the junction voltage, $k = 1.38 \times 10^{-23}$ J/deg Kelvin is Boltzmann's constant, and T is temperature in degrees Kelvin. Figure 5.2(b) plots (5.1) for a typical diode, with the reverse bias and forward bias regions of operation delineated.

The forward bias region is sometimes termed the *square-law* region because the diode current is roughly proportional to the square of the junction voltage. More generally, we can expand (5.1) in the power series form [2]

$$I = aV_J + bV_J^2 + cV_J^3 + \dots \tag{5.2}$$

If V_J is a microwave signal with a sinusoidal frequency variation of the form $V_{RF}\cos\omega t$, then (5.2) will yield a diode current having components with frequencies that are harmonics of the fundamental frequency $\omega = 2\pi f$:

$$I = (b/2)V_{RF}^2 + \left[aV_{RF} + (3c/4)V_{RF}^3\right]\cos\omega t \tag{5.3}$$
$$+ (b/2)V_{RF}^2 \cos 2\omega t + (c/4)V_{RF}^3 \cos 3\omega t$$

where we have assumed for simplicity that the coefficients of the voltage terms that are greater than third order in (5.2) are zero. The second-order (2ω) and third-order (3ω) harmonics of the fundamental frequency that appear in (5.3) are the basis for frequency multiplication. The presence of a DC term means that a diode can be used to convert an RF voltage to a DC current, the essence of detection.

A diode's ability to mix signal frequencies can be demonstrated in a similar manner if we combine a local oscillator (LO) signal and a signal at an intermediate frequency (IF) at the diode junction: $V_J = V_{LO} \cos \omega_{LO} t + V_{IF} \cos \omega_{IF} t$. If we substitute this expression into (5.2), the voltage-squared term gives, along with DC and fundamental frequency signals, the standard mixer RF output, consisting of signals at frequencies that are the sum and difference of the LO and IF frequencies [2].

The *pn* diode has relatively high capacitance, and thus it is slow to respond to microwave signals. Microwave circuit engineers often use the Schottky metal-semiconductor diode shown in Figure 5.3. Unlike the *pn* junction, the Schottky junction comprises just one semiconductor, of type n, abutted to a metal. Because the metal's free-electron concentration is orders of magnitude higher than that of the semiconductor, the depletion region extends almost entirely into the semiconductor. The junction capacitance is low, and with only electrons as charge carriers, the Schottky diode is capable of much higher frequency operation than the *pn* diode.

The active region comprising the conductor and semiconductor material of a typical microwave diode is extremely small and fragile. Consequently, diodes are often packaged for use in microwave circuits. Figure 5.4(a) shows a diode chip mounted in a typical cylindrical metal package. The cathode is bonded to a pedestal, and gold ribbon is used to attach the anode to the other end cap of the package [3]. The package is so small that its electrical characteristics can be represented by lumped elements at the frequencies of most applications. These elements are the package capacitance, C_P, and the gold ribbon's inductance, L_S, and resistance, R_S. A diode's equivalent circuit schematics for forward and reverse bias appear in Figure 5.4(b, c) with the package parasitic elements included.

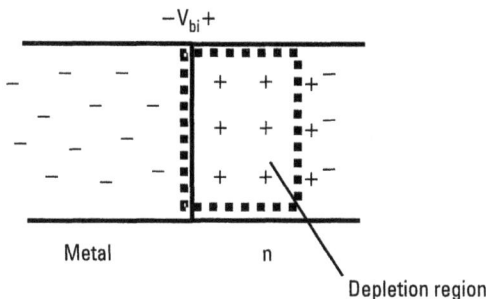

Figure 5.3 Schottky junction. High free-electron concentration in metal causes the depletion region to extend almost entirely into semiconductor.

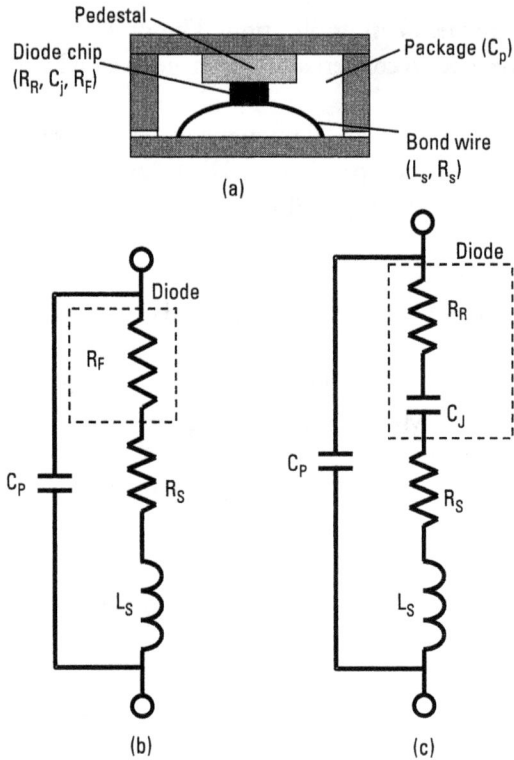

Figure 5.4 (a) Packaged diode chip. Packaged diode models: (b) forward biased and (c) reverse biased.

For a forward biased diode having the equivalent circuit shown in Figure 5.4(b), the diode impedance Z_D is given by

$$Z_D = Z_{FB} = (1/j\omega C_P) // (R_F + R_S + j\omega L_S)$$ (5.4)

where the symbol // represents the parallel combination of the two expressions on either side. Similarly, for the reverse biased diode as represented by Figure 5.4(c), we have

$$Z_D = Z_{RB} = (1/j\omega C_P) // (R_R + 1/j\omega C_J + R_S + j\omega L_S)$$ (5.5)

5.2.2 Varactors

A varactor is a reverse biased diode that is used as a voltage-variable capacitor as shown in Figure 5.5. Varactors are used to tune reactively other devices, such as the negative resistance device oscillator represented schematically in Figure 5.6.

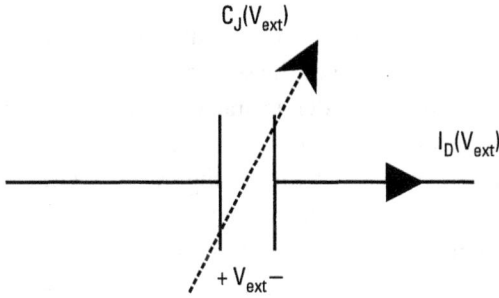

Figure 5.5 A varactor is a reverse biased diode that is used as a voltage-variable capacitor.

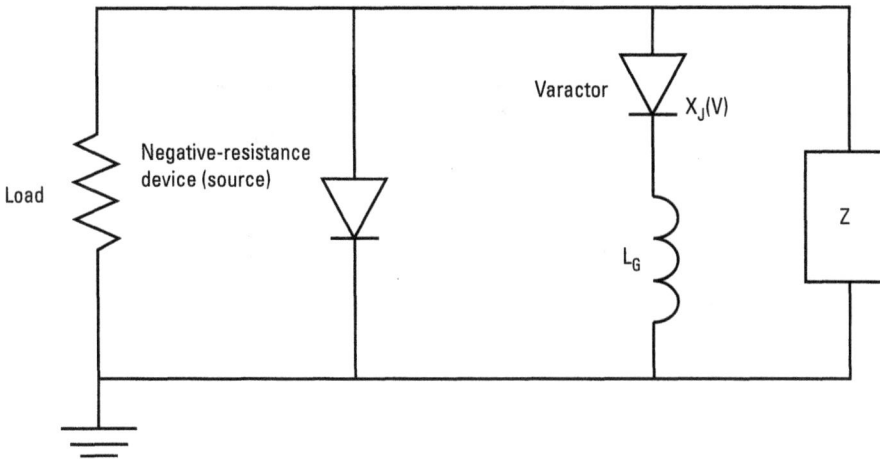

Figure 5.6 A negative-resistance diode oscillator can be voltage tuned in frequency using a varactor. Inductance L_G between the varactor mount and ground can perturb the frequency tuning range.

The capacitance of the varactor can be adjusted with reverse bias voltage so that at a particular frequency f_0 the negative resistance diode's reactance, $X_D(f_0)$, is equal to the negative of the reactance of the remainder of the embedding circuit including the varactor, $X_V (f_0, V)$:

$$X_D(f_0) + X_V(f_0) = 0$$

If the resistances of the negative resistance device and the embedding circuit sum to zero at f_0 also, the device will oscillate [4]. The oscillator is *voltage controlled* if we can vary the oscillation frequency by changing the varactor's reverse bias voltage. The range of frequencies over which the oscillator can be made to operate is called the tuning range, and it depends on the varactor's range of reactance. The varactor's grounding is important because any inductance between the varactor and ground will change its reactance along with the frequency tuning range of the oscillator.

As a simple example, consider a varactor diode that has a reverse bias junction capacitance range of 0.4 to 2 pF. The other parameters in the diode's equivalent circuit are listed in the caption of Figure 5.7. The varactor is part of a voltage-controlled oscillator (VCO) that tunes from 5.5 to 11.0 GHz. From (5.5), we plot the varactor's reactance curve for zero ground impedance ($L_G = 0$) in Figure 5.7(a). The varactor's reactance at 5.5 GHz is –10 ohms, and at 11.0 GHz it is –25 ohms. The figure also plots the negative resistance device's reactance curve, $-X_D(f)$. If we assume that these two curves are computed at the same reference plane within the oscillator circuit, then at frequencies where they

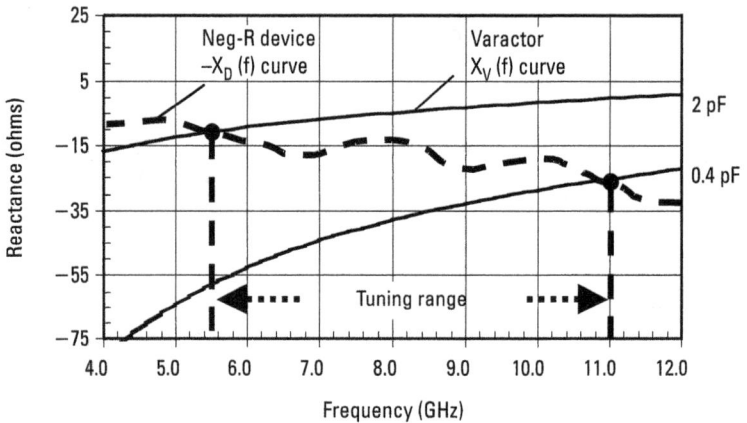

(a)

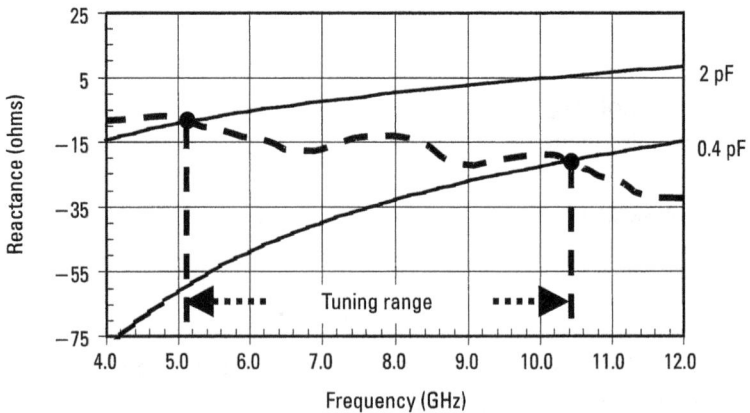

(b)

Figure 5.7   VCO tuning range shifts down in frequency with ground inductance: (a) $L_G = 0$ nH; and (b) $L_G = 0.1$ nH. Other parameters: $R_R = 0.5 \ \Omega$, $C_P = 0.08$ pF, $L_S = 0.1$ nH, $R_S = 0.2 \ \Omega$, $R_G = 0 \ \Omega$.

cross, oscillation will occur (assuming the corresponding resistance curves cross also). If we add just 0.1 nH of ground inductance, the reactance curve shifts up and the tuning range shifts down almost 500 MHz as shown in Figure 5.7(b).

5.2.3 Diode Limiters and Switches

Unlike a varactor, a microwave switch uses both the forward and reverse biased states of a diode. A common switch configuration, shown in Figure 5.8(a), mounts a diode in shunt across a transmission line. When the diode is reverse biased, it acts as an open circuit, and the transmission line propagates energy with low loss. Conversely, when the diode is forward biased, it acts as a short circuit and reflects incident waves propagating on the transmission line.

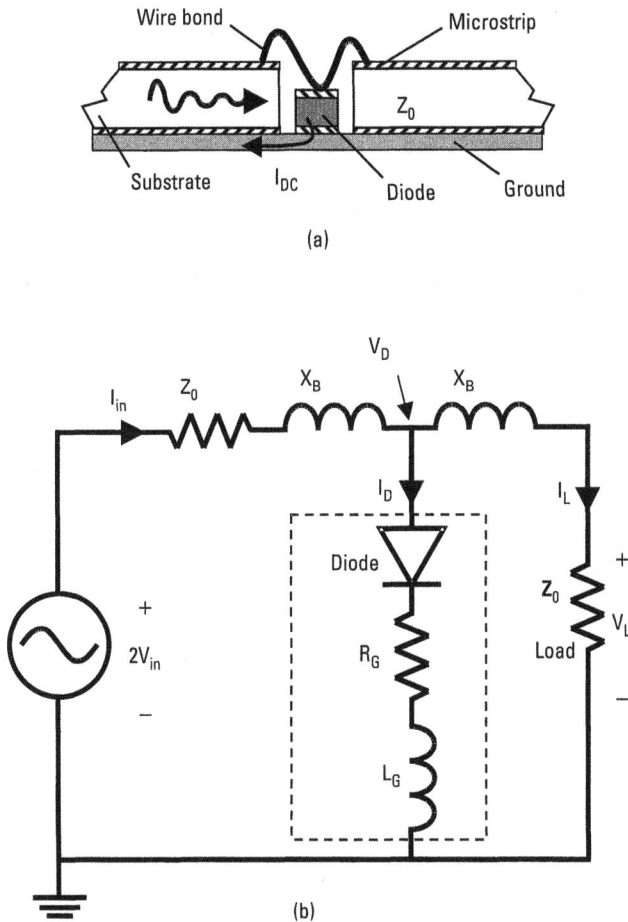

(a)

(b)

Figure 5.8 (a) Packaged diode mounted in shunt with a microstrip transmission line. (b) Circuit schematic.

If the diode is unbiased, it will act as a power *limiter*. When a relatively low level of microwave power is incident, the DC current term in (5.3) is small, and the junction voltage is just the built-in voltage. In this state, the diode is capacitive, and the limiter transmits the incident signal. As the incident microwave power level increases, the diode develops an increasingly large DC voltage, which forward biases the junction and shorts the transmission line. Above a certain power level, all incident signals are increasingly attenuated by the limiting action.

The operation of a shunt diode switch is similar, except the applied DC bias places the diode either into the forward bias region (off state) or the reverse bias region (on state). The bias is set to overcome any DC voltage the diode might generate by rectifying the incident microwave power. In the on state, the switch has low insertion loss; while in the off state, it has high isolation.

The insertion loss and isolation are ideally zero and infinite, but they are limited by the diode's characteristics, including its reverse bias capacitance, forward resistance, and the package parasitics. These are usually reduced as much as possible to provide adequate performance at the frequency of operation. Additional inductance and resistance (L_G, R_G) are created in the process of mounting the diode in the circuit. Both the diode package bond wire (L_S, R_S) and package mount (L_G, R_G) contribute to ground path impedance, and they degrade the operation of a switch.

The degradation caused by resistance and inductance in the ground path of a shunt diode can be appreciated by applying some simple circuit analysis to the switch circuit shown in Figure 5.8. The diode is embedded in a transmission line with characteristic impedance Z_0, and an incident TEM wave from the left. Following the approach of Bahl and Bhartia [5], we define the switch's insertion loss as the ratio of the power delivered to the transmission line on the right (the load) by an ideal switch in the on state with perfect grounding ($R_G = 0$, $L_G = 0$) to the actual power delivered by the real switch. By ideal switch, we mean a switch with infinite impedance in the reverse bias or on state, and zero bond wire reactance, $X_B = 0$. In this case, exactly half the source voltage (V_{in}) will fall across the load. If V_L is the voltage across the load when the real switch is in place, then the insertion loss is given by

$$\text{Insertion loss} = \left| V_{in} / V_L \right|^2 \tag{5.6}$$

where $V_L = I_L Z_0$.

The current I_{in} flowing from the source is given by

$$I_{in} = I_D + I_L = V_D / Z_D + V_D / (X_B + Z_0) = V_D \left[1/Z_D + 1/(X_B + Z_0) \right] \tag{5.7}$$

where Z_D is the diode impedance [from (5.4) or (5.5)] plus the ground impedance, $R_G + L_G$. The diode voltage V_D is found from

$$
\begin{aligned}
V_D &= 2V_{in} - I_{in}(X_B + Z_0) \\
&= 2V_{in} - V_D[1/Z_D + 1/(X_B + Z_0)](X_B + Z_0) \\
&= 2V_{in} - V_D(X_B + Z_0 + Z_D)/Z_D
\end{aligned}
\tag{5.8}
$$

If we solve (5.8) for the diode voltage, we get

$$
V_D = 2V_{in} Z_D/(X_B + Z_0 + 2Z_D)
\tag{5.9}
$$

We can now write an equation for the load voltage using (5.9):

$$
\begin{aligned}
V_L &= I_L Z_0 \\
&= V_D Z_0/(X_B + Z_0) \\
&= 2V_{in} Z_D Z_0/[(X_B + Z_0)(X_B + Z_0 + 2Z_D)]
\end{aligned}
\tag{5.10}
$$

which we solve for the insertion loss of the switch:

$$
\begin{aligned}
\text{Insertion loss} &= |V_{in}/V_L|^2 \\
&= |(X_B + Z_0)(X_B + Z_0 + 2Z_D)/(2Z_D Z_0)|^2
\end{aligned}
\tag{5.11}
$$

With (5.11), we can study the effects of imperfect grounding on switch
performance. Figure 5.9(a) plots the insertion loss versus frequency of a shunt
diode switch in the on state for various values of ground resistance and inductance. While ground resistance has a small effect on insertion loss, ground inductance increases insertion loss significantly, particularly at higher frequencies.

Figure 5.9(b) plots the isolation of the switch in the off state. Since isolation depends on the diode acting as a short circuit, both ground resistance and
inductance can seriously decrease the isolation of the switch. Half an ohm of
ground resistance reduces the isolation by 7 dB. Isolation is severely affected by
even a small amount of ground inductance. In this example, just 0.1 nH reduces
the isolation by 10 dB above 2 GHz. Consequently, a low impedance direct connection of the package to ground as in Figure 5.8(a) is highly desirable—only the
package's internal bond wire impedance limits the switch's performance. If the
switch must be mounted to the microstrip layer of a circuit board rather than
directly to its RF ground plane, an array of vias to ground should be used to minimize the inductance, a subject we discuss in Section 5.5.2.

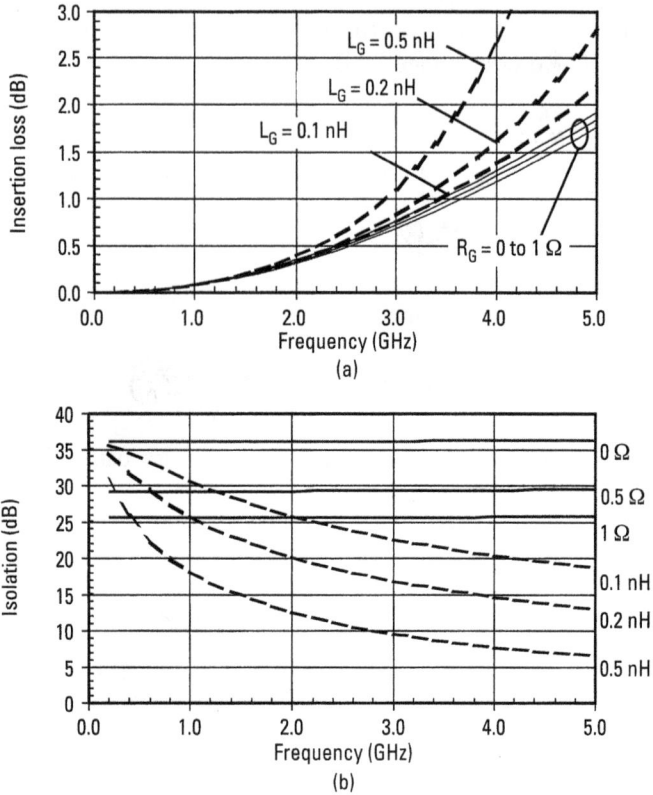

Figure 5.9 Shunt diode performance as a function of ground inductance and resistance: (a) insertion loss in on state; and (b) isolation in off state. $C_J = 1$ pF, $R_F = 0.4$ ohms, $R_R = 0.5$ ohms, $C_P = 0.08$ pF, $L_B = 0.3$ nH; $L_S = 0$ nH, $R_S = 0$ ohms, $Z_0 = 50$ ohms.

5.2.4   Diode Mixers

A forward biased diode produces a current that is rich in harmonics of the fundamental frequency. If we drive a Schottky diode as in Figure 5.10 with two signals at different frequencies (LO and IF), a variety of signals will appear at the RF port with frequencies at different combinations of the LO and IF frequencies. A properly designed mixer has filters at its LO, IF, and RF ports that permit only the desired signal(s) to appear at the RF port. In addition, the diodes in a mixer frequently require a DC connection to ground, since the mixing process usually generates a DC current. This same ground will be the reference for the microwave signals also. The ground path needs to have low inductance and also must be well characterized over the IF, LO, and RF frequency ranges so that the mixer design can be compensated for its influence.

As an example, Figure 5.11 shows a subharmonic diode mixer implemented with microstrip transmission line. The mixer up-converts an IF signal at

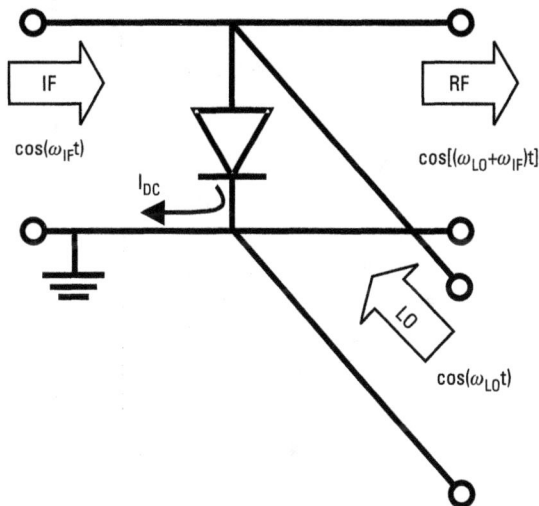

Figure 5.10 Simplified diode mixer (upconverter) schematic. (*After:* [4].)

a frequency of 5.3 GHz to an RF frequency of 24.2 GHz [private communication with S. R. Ramirez]. *Subharmonic* means that the LO port is driven at a subharmonic, 9.45 GHz, of the LO frequency at 18.9 GHz. Mixing is accomplished with a pair of antiparallel diodes, a common configuration for subharmonic mixers, which outputs an odd-order mixing product, $2\omega_{LO} + \omega_{IF}$ [6]. If the diodes are perfectly matched in their electrical characteristics, no DC current will flow. In practice, some DC current flows, so a ground path is required. In this mixer, the vias in the LO filter provide the connection to the ground plane, and they also serve to reject the RF signal when the mixer is used as a downconverter. Since the vias affect the performance of the mixer from DC to RF (24 GHz) and higher, it is important that they have very low inductance. Even so, the vias' electrical characteristics must be included in the mixer circuit model and quantified very accurately across a wide frequency band, ideally by numerical electromagnetics analysis software.

The mixer in Figure 5.11 was designed using a nonlinear circuit simulator. We can appreciate the influence of grounding on its performance by perturbing the via grounding and rerunning the simulator. Figure 5.12 shows the simulator's predicted IF-to-RF conversion loss and IF port-return loss as a function of the IF input power level for different amounts of via inductance. As a basis for comparison, we simulated the mixer with perfect vias having no inductance. The IF-to-RF conversion loss is a minimum for an IF power level of −7 dBm. If we increase the via inductance to 0.1 nH, the conversion loss increases about 1.5 dB, and the optimum IF power increases to −4 dBm. Increasing inductance another 0.1 nH, adds another 0.7 dB to conversion loss and shifts the optimum

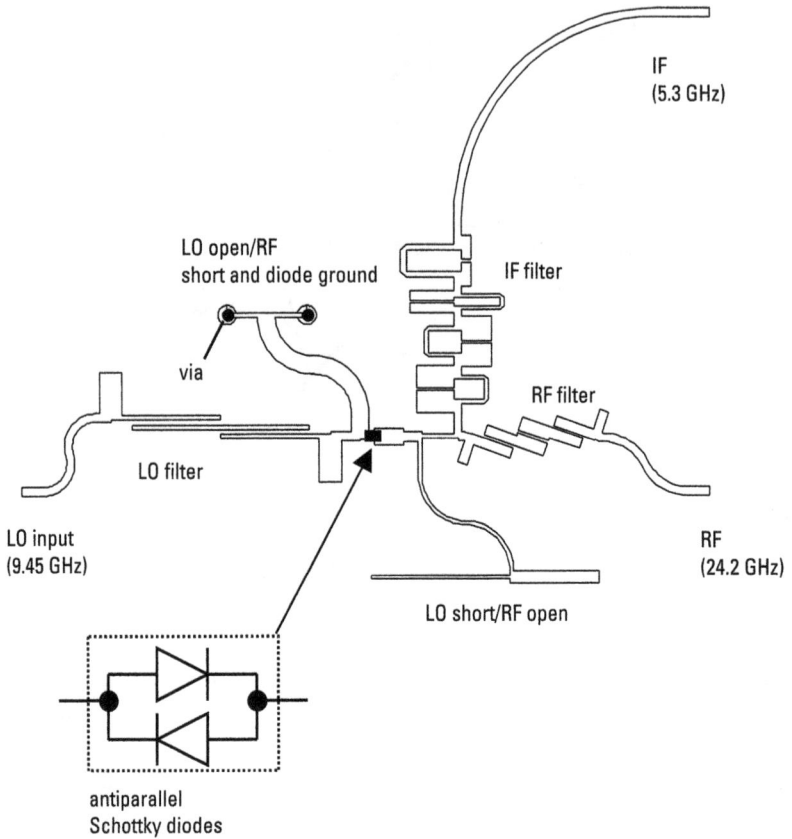

Figure 5.11 Subharmonic diode mixer layout for microstrip implementation. [*Source:* Private communication with S. R. Ramirez.]

IF power up another decibel. If the designer knows the via's inductance precisely, he can adjust the mixer design to give optimum performance for that inductance..

5.3 Microwave Transistors

Microwave transistors have become ubiquitous in solid-state based microwave modules and systems. Their small size, low cost, and compatibility with printed circuit transmission lines such as microstrip give them a tremendous advantage in most applications over tube-based technology. Transistors fall into two main categories: *bipolar junction transistors* (BJTs) and *field effect transistors*. GaAs metal semiconductor FETs (MESFETs) operate to much higher frequencies than BJTs and find applications in *planar microwave circuits*, which include planar transmission lines such as microstrip that provide the matching circuitry

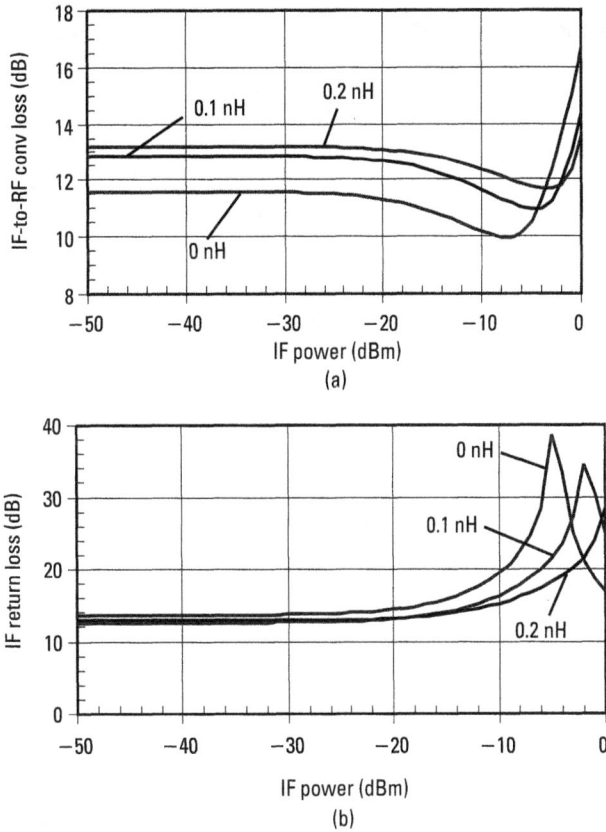

Figure 5.12 Subharmonic diode mixer (a) IF-to-RF conversion loss and (b) IF return loss as a function of ground return inductance.

necessary to optimize a MESFET for operation as an amplifier, oscillator, or other circuit.

Planar circuits are constructed in two ways. A *hybrid* or *microwave integrated circuit* (MIC) includes one or more unmatched MESFETs or diodes along with discrete, separately processed components, including transmission lines, inductors, capacitors, and resistors. These components are interconnected using bond wires or metal ribbons, solder, and epoxy. Although the MESFET is very small, the other components in a MIC, particularly the transmission lines, are relatively large, with areas measured in square inches. When large production volumes are required, a planar circuit usually is realized as a *monolithic microwave integrated circuit* (MMIC), a single piece of semiconductor that often includes all the components required for a fully operational microwave circuit, namely the transistors, transmission lines, capacitors, inductors, resistors, and their interconnections. MMICs are far smaller than hybrid circuits, with areas measured in square millimeters rather than square inches. The capability to batch process thousands of MMICs on a single wafer of semiconductor has led to their extremely low cost. Further, the introduction of high electron mobility transistor (HEMT) technology in the 1990s has enhanced the performance of the MESFET MMICs and enabled them to operate well into the millimeter-wave region. We will focus on grounding techniques for MESFETs, but most of the concepts we discuss apply equally well to BJTs.

5.3.1 Operational Fundamentals

The MESFET, shown in Figure 5.13, is a three-terminal, voltage-controlled current source. Current (electrons) flows between the source and drain through a layer of semiconductor called the channel. Frequently, the drain is DC biased a few volts positive, and the source is grounded so that current flows from the drain to the source. The gate metal forms a Schottky diode junction with the channel semiconductor. A voltage applied at the gate, which determines how far the diode's depletion region extends into the channel, regulates the amount of drain current that flows. Most MESFETs are designed so the maximum drain current flows when the gate bias is near zero volts, and the depletion region width is a minimum. As the gate is reverse biased, the depletion region narrows the channel until it is pinched off, and no drain current can flow. The Schottky junction has very low capacitance and can respond to an AC voltage at the gate that is modulated at microwave frequencies. The MESFET amplifies the AC voltage by replicating the modulating waveform's characteristics in the drain current. A high performance MESFET with input and output matching networks can achieve 20 dB of amplification or *gain*.

MESFETs are used most frequently as microwave amplifiers in circuits such as low noise receivers, high power transmitters, and interstage gain blocks. Figure 5.14 shows a simplified, single-stage MESFET amplifier schematic. A

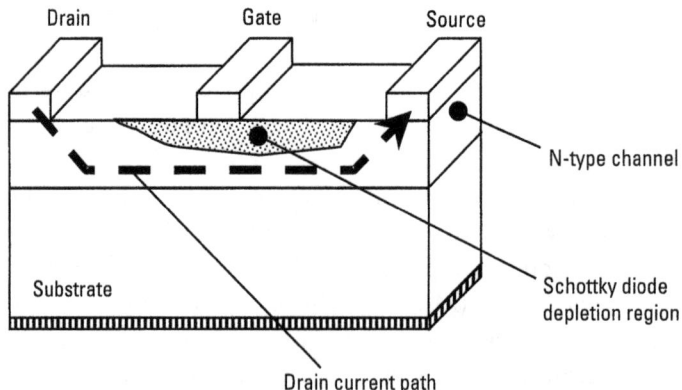

Figure 5.13 MESFET semiconductor structure. (*After:* [1], pp. 160–166.)

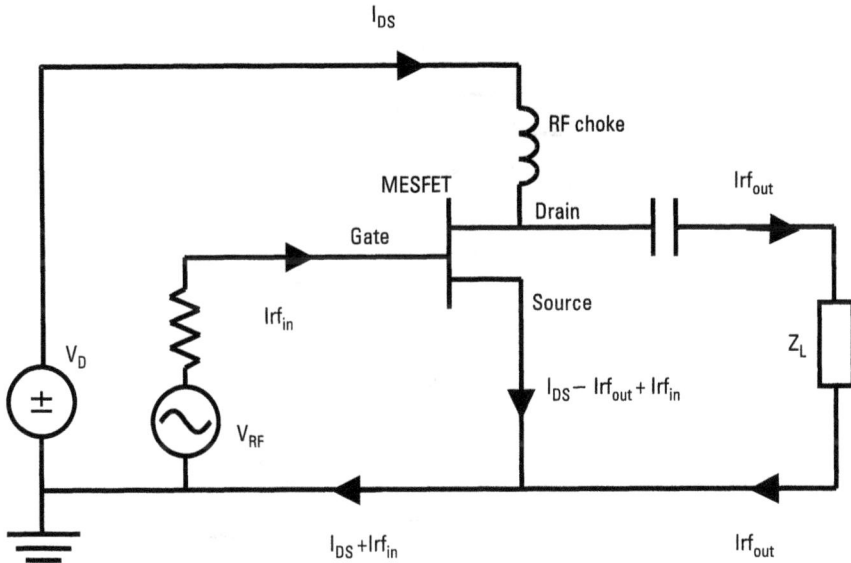

Figure 5.14 MESFET RF amplifier circuit schematic.

DC source V_D establishes the drain voltage relative to the MESFET's grounded source. An RF source V_{RF}, which might be a previous amplifier stage or an oscillator, provides the input RF signal at the MESFET gate. A portion of the drain current, modulated at the frequency of the gate signal, flows to the load. As the schematic shows, both DC and RF ground currents are flowing, often following the same conducting path, back to their respective sources.

5.3.2 Source Resistance and DC Grounding

For many amplifiers, the MESFET's source contact provides the ground reference for the device. Poor grounding of the source can lead to serious performance problems. Figure 5.15(a) shows a MESFET with resistance in the path to ground. Because the DC voltage V_D is often fixed, drain current flowing through the resistance in the source increases the source voltage and reduces V_{DS} across the MESFET as shown in Figure 5.15(b). For zero source resistance, the AC signal amplitude can swing from V_{min} to V_{max}. If there is source resistance, the voltage swing is reduced to V_{min} to V_0. The larger the drain current is, the greater the reduction in V_{DS} for a given source resistance. Thus, high power devices, which may draw several amps of drain current, require extremely low resistance grounding.

    As an example, consider the FL3135-18F internally matched power FET produced by Eudyna (formerly Fujitsu Compound Semiconductor) [7]. This device generates 20W of microwave power from 3.1 to 3.5 GHz with 10.5-dB

(a)

(b)

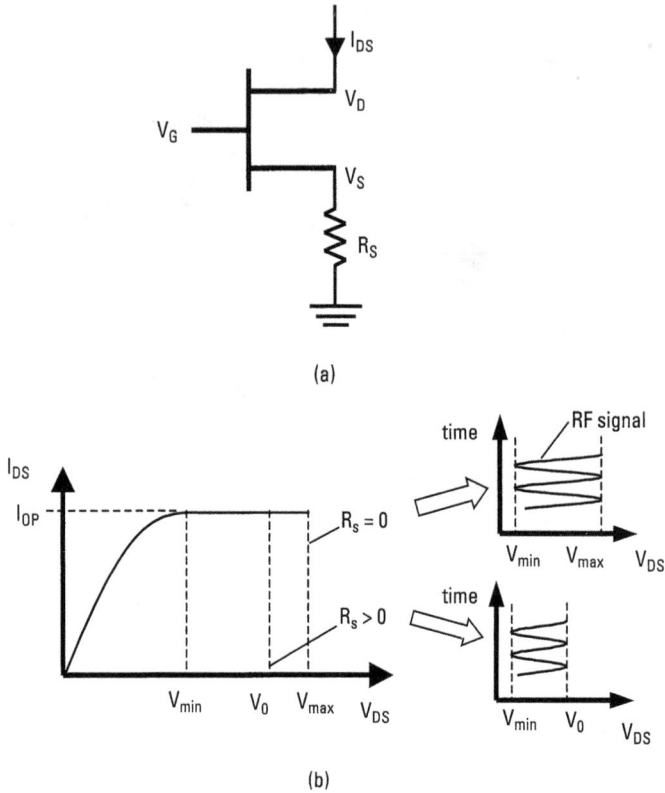

Figure 5.15 (a) Source resistance R_s in a MESFET's ground path (b) reduces the output RF voltage range.

gain. The FET's drain DC bias is 10 volts with 4.8 amps of drain current. The *power-added efficiency* is calculated from

$$\eta_{pa} = \left(P_{out} - P_{in} \right) / P_{DC} \qquad (5.12)$$

where $P_{in} = P_{out}/\text{Gain} = 20\text{W}/11.2 = 1.8\text{W}$, and the efficiency is $(20\text{W} - 1.8\text{W})/48\text{W} = 38\%$ if the source resistance is 0 ohms. If we solve (5.12) for P_{out}, we get

$$P_{out} = P_{in} + \eta_{pa} P_{DC} \qquad (5.13)$$

If the DC bias is held fixed at 10 volts, and we introduce just 0.1 ohms of source resistance, the source voltage rises to $(4.8\text{A})(0.1 \text{ ohms}) = 0.48\text{V}$, and V_{DS} across the device drops from 10V to 9.52V. Such a reduction in drain-to-source voltage, if left uncompensated, can compress the range over which the output microwave signal can vary [see Figure 5.15(b)]. Decreased gain, output power, and efficiency, and increased distortion may result. For the Eudyna power amplifier, assuming P_{in} and efficiency are unchanged, $P_{DC} = (9.5\text{V})(4.8\text{A}) = 45.6\text{W}$, and we get $P_{out} = 19.1\text{W}$ from (5.13). Thus, a mere 0.1 ohms of source resistance causes a 5% drop in output power, and DC grounding has directly influenced RF performance. One way to compensate for source resistance is to measure the drain-to-source voltage and increase V_D by 0.5 to 10.5 volts. If a voltage regulator biases the amplifier, the regulator must have sufficient voltage headroom. If the regulator takes its input from a 12-volt DC-to-DC converter, such headroom may not be available.

5.3.3 Source Inductance, Feedback, and RF Grounding

Although source resistance can degrade the performance of an amplifier, inductance in the source to ground path is a far more serious problem, because it can turn an otherwise unconditionally stable amplifier into an uncontrollable oscillator. The oscillations are likely to cause the system to malfunction, and they may cause the amplifier to burn out.

We can understand better why source inductance causes an amplifier to oscillate if we analyze the MESFET equivalent circuit shown in Figure 5.16. The current I_g flows into the gate junction from the left. The junction is reverse biased during normal operation, so it is represented by a variable capacitance having a voltage V_{gs}. The drain current dependence on the gate voltage is represented by the current source $g_m V_{gs}$, where g_m, the transconductance, is related to the MESFET's gain. Z_o is the complex output impedance of the MESFET. The impedance Z_s includes any impedance in the source connection to ground. If $Z_s = 0$, the source is grounded, the source voltage $V_s = 0$ volts, and the gate and drain currents are single point grounded (compare with Figure 1.4). Then, the only dependence between the drain and gate currents is the unidirectional one established by $g_m V_{gs}$.

On the other hand, if Z_s is not zero, V_s is not at ground, and the grounding of I_g and I_d has the multipoint configuration (see Figure 1.3). Both the drain and gate currents use the path through Z_s as a common ground connection. The voltage V_s is dependent on the drain current, causing the input voltage V_{gs} to become dependent on the drain current also. Thus, a nonzero source impedance reverse couples the MESFET's output to its input, which is known as a *feedback* condition. The series feedback shown in Figure 5.16 will detune the amplifier and can lead to oscillation, as we shall now show more methodically.

We can analyze the equivalent circuit in Figure 5.16 as a two-port circuit. The input port is the gate terminal referenced to ground, and the output port is the drain referenced also to ground. The scattering parameters for a two-port circuit are given by (3.7) and (3.8) from Chapter 3, which we rewrite here:

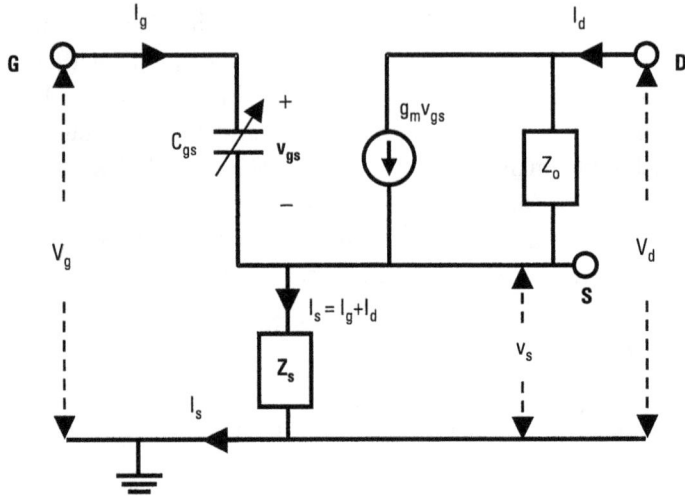

Figure 5.16 MESFET equivalent circuit schematic.

$$b_1 = S_{11}a_1 + S_{12}a_2 \tag{5.14}$$

$$b_2 = S_{21}a_1 + S_{22}a_2 \tag{5.15}$$

For the MESFET, S_{11} is the gate input match, S_{22} is the drain output match, S_{21} is the voltage gain, and S_{12} is the reverse isolation. If the source impedance is 0 ohms, S_{12} is equal to zero; otherwise, it is not zero. We will derive an expression for S_{12} to prove this statement.

Since the equivalent circuit parameters are voltages and currents, it is sensible to start with Y-parameters [8], and in particular, derive

$$Y_{12} = I_g / V_d \text{ with } V_g = 0 \tag{5.16}$$

We will relate S_{12} and Y_{12} shortly. From Figure 5.16, we can write

$$V_g = v_{gs} + I_s Z_s = I_g X_{gs} + I_s Z_s \tag{5.17}$$

where $X_{gs} = 1/j\omega C_{gs}$. Now, Y_{12} is defined for $V_g = 0$, for which (5.17) gives

$$I_s = -I_g X_{gs} / Z_s \tag{5.18}$$

Alternatively, we write the source current as

$$I_s = I_g + g_m v_{gs} + (V_d - V_s)/Z_o \tag{5.19}$$

If we equate (5.18) and (5.19), and use $V_s = I_s Z_s$ and $v_{gs} = I_g X_{gs}$, we get

$$-I_g X_{gs}/Z_s = I_g + g_m I_g X_{gs} + V_d/Z_o + I_g X_{gs}/Z_o \qquad (5.20)$$

Finally, we solve (5.20) for Y_{12}:

$$Y_{12} = \frac{I_g}{V_d}\bigg|_{V_g=0} = -\frac{1}{Z_0} \cfrac{1}{1 + \cfrac{1}{j\omega C_{gs}}\left(g_m + \cfrac{1}{Z_s} + \cfrac{1}{Z_o}\right)} \qquad (5.21)$$

It can be shown that the reverse isolation is given in terms of Y_{12} as [9]

$$S_{12} = \frac{-2Y_{12}}{(Y_{11}+1)(Y_{22}+1) - Y_{12}Y_{21}} \qquad (5.22)$$

If $Z_s = R_s + j\omega L_s = 0$, then the denominator of Y_{12} is infinite, and Y_{12} is equal to zero. As long as both Y_{11} and Y_{22} are not -1, it follows that if $Y_{12} = 0$, $S_{12} = 0$, and there is no feedback.

An unstable amplifier oscillates, and a *conditionally stable* amplifier oscillates when specific impedances are presented at its input and/or output ports. An amplifier usually is designed to be *unconditionally stable*, meaning that it cannot be made to oscillate no matter what impedances terminate its ports. The conditions for unconditional stability are met if an amplifier's *S*-parameters satisfy [10]

$$K = \frac{1 - |S_{11}|^2 - |S_{22}|^2 + |\Delta|^2}{2|S_{12}S_{21}|} > 1 \qquad (5.23)$$

and

$$|\Delta| = |S_{11}S_{22} - S_{12}S_{21}| < 1 \qquad (5.24)$$

where K is called the *Rowlett stability factor*. The importance of low source impedance is implicit in the $S_{12}S_{21}$ factor appearing in the stability conditions. Since $|S_{21}|$ can be well above unity for a high gain amplifier, it is important to keep S_{12}, which depends on source impedance, small so that (5.23) and (5.24) are satisfied. If $S_{12} = 0$, and if we assume that $|S_{11}|$ and $|S_{22}|$ are less than unity, then K will be infinite, and $|\Delta| < 1$. As the gain of an amplifier increases, low source impedance becomes essential for keeping S_{12} as small as possible.

When an engineer designs a MMIC amplifier using circuit analysis software, he often leaves to someone else the problem of grounding the MMIC to the circuit board. As many as three different engineers may be involved in designing the entire path to ground: (1) the MMIC designer, (2) the MMIC package designer, and (3) the circuit board designer. In effect, the MMIC designer neglects the ground path from the MMIC to the DC supply by assuming it has zero impedance. Thus, the S-parameters supplied with a commercially available MMIC amplifier probably do not include the effects of the entire grounding path. An engineer designing a microwave circuit board with such a MMIC must make sure the characteristics of the grounding path are adequate. Otherwise, when he places the MMIC into his circuit with imperfect grounding, the magnitude of S_{12} may increase and destabilize the amplifier.

Besides instability, source inductance degrades amplifier gain, often by several decibels [11]. A simple expression approximates the maximum available gain for a MESFET having the equivalent circuit of Figure 5.16 [12]:

$$G_{max} = \frac{\left(f_T\big/f\right)^2}{\dfrac{4}{R_o}\left(R_s + \pi f_T L_s\right)} \tag{5.25}$$

where $f_T = g_m\big/2\pi C_{gs}$, R_o is the resistive part of Z_o, and R_s and L_s comprise Z_s. Equation (5.25) assumes the device is unconditionally stable. For low noise amplifiers, source inductance increases noise figure also, as discussed by Vendelin [13].

5.3.4 Examples

As an example, consider the single-stage MESFET amplifier in Figure 5.17. A packaged MESFET is mounted on the RF conductor layer of a circuit board. Microstrip transmission lines are connected to the gate (input) and drain (output) leads of the MESFET, and its source leads are connected to the circuit board ground with via holes. The transmission line matching circuits at the input and output have been designed for gain and input and output match over the 4- to 7-GHz band.

Figure 5.18 plots the gain and stability factor of the amplifier for different amounts of source inductance. For perfect grounding (0 nH), the amplifier's gain is maximized around 4.2 GHz, after which it gradually falls off with increasing frequency. The stability factor is greater than one from 2 to 8 GHz, and it is rising at the band edges: the amplifier is unconditionally stable. With 0.1 nH of source inductance, the gain has decreased 1.5 dB at 4.2 GHz, and it

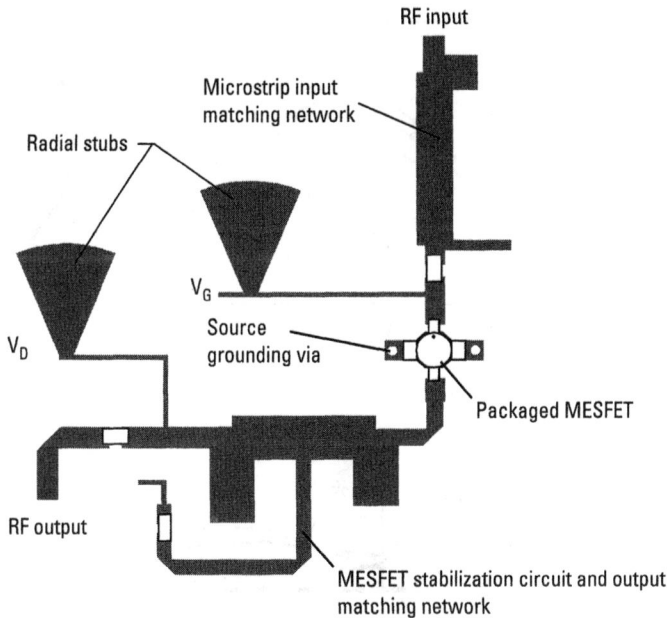

Figure 5.17 Single-stage microstrip MESFET amplifier layout.

has increased slightly between 7 and 8 GHz. The stability factor, while still greater than one, has decreased, particularly between 7 and 8 GHz. When we increase the source inductance to 0.25 nH, the stability factor falls below one around 3 GHz and above 6 GHz. The amplifier's gain has increased at the high end of the frequency band—there is a pronounced peak at 7.3 GHz, which is an indication that the amplifier is likely to oscillate. The gain at 4.2 GHz has fallen another 1.5 dB. For this amplifier to perform near its optimum level, we should add more source vias to the source grounding traces shown in Figure 5.17. Recall from our discussion in Section 3.6 on via hole modeling that 0.1 nH is approximately the inductance we can expect from two adjacent vias, and 0.25 nH is the inductance of a single via. Thus, we might want to increase the number of vias grounding each of the MESFET's source leads to two or more. In addition, one should place the vias as close as possible to the edge of the MESFET package and use a thin substrate to minimize via inductance as suggested by Wei et al. [14]. Finally, every amplifier circuit model should include an accurate model of the via inductance so that the amplifier's stability can be predicted, and an effective stabilization circuit (see Figure 5.17) can be designed if necessary.

We have concentrated on amplifiers, but MESFETs can perform a variety of circuit functions. Low noise fixed frequency and voltage controlled oscillators often employ MESFETs as the negative resistance device. In contrast to an amplifier, a MESFET oscillator needs a feedback element such as a capacitor in

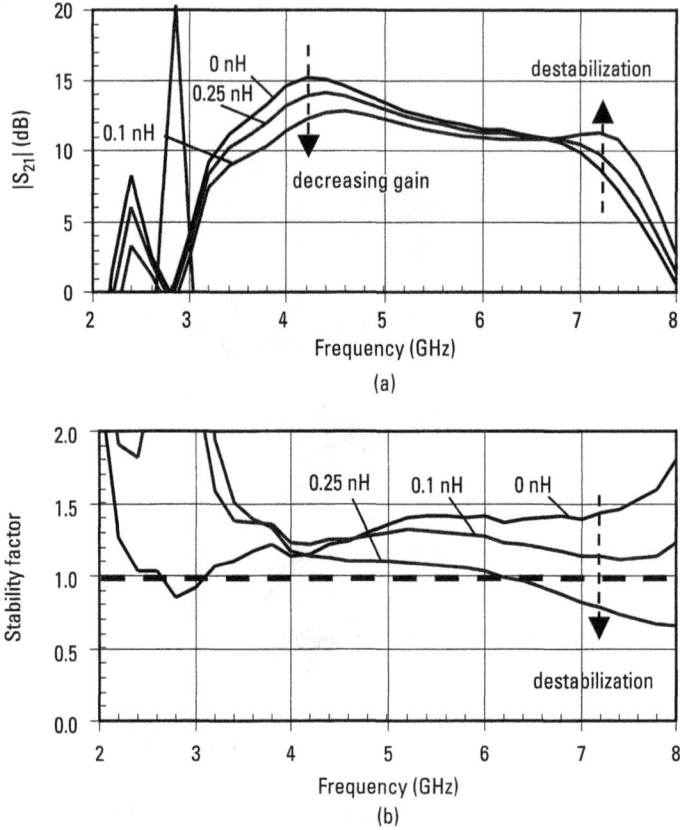

Figure 5.18 MESFET amplifier (a) forward transmission and (b) stability factor K reveal decreasing stability as source inductance increases. Device: Sirenza SPF-2086T(K), $V_{DS} = 5V$, $I_{DS} = 40$ mA. (*After:* [15].)

the source to ground path as shown in Figure 5.19. As with the diode oscillator we investigated in Section 5.2.2, inductance in the source to ground path will perturb a MESFET oscillator's behavior. For example, consider a fixed frequency oscillator of the type shown in Figure 5.19, with $C_S = 1$ pF for oscillation at 16 GHz. If we neglect the source inductance and resistance, then the reactance of the capacitor at 16 GHz is $1/j\omega C_S = -j/(2\pi\ 16 \times 10^9 \times 10^{-12}) = -j10$ ohms. If the source inductance is 0.1 nH, then the source reactance at 16 GHz due to the capacitor plus inductance is given by $1/j\omega C_S + j\omega L_S = -j10$ ohms $+ j2\pi \times 16 \times 10^9 \times 10^{-10}$ ohms $= -j10$ ohms $+ j10$ ohms $= 0$ ohms. Most likely, this change in source reactance will cause the oscillator to change frequencies. Thus, a precise model of the ground path is essential for accurately predicting the oscillation frequency.

Figure 5.20 shows another oscillator example, a 20-GHz VCO circuit that is among the example files included with Agilent Technology's *Advanced Design*

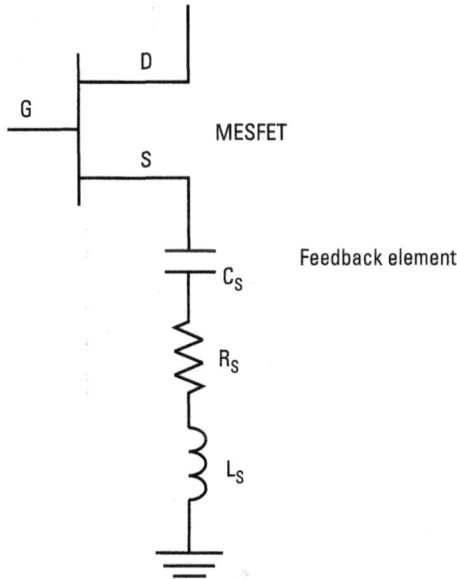

Figure 5.19 Ground path source resistance R_S and inductance L_S can perturb the operating point of a transistor oscillator by modifying the feedback reactance.

System (ADS) software [16]. The nominal design, which includes detailed ground path modeling, can be tuned in ADS from 19.32 to 19.85 GHz as shown by the upper curve plotted in Figure 5.21. If we add just 0.1 nH to the gate and drain bias connections ($L_G = 0.1$ nH), the entire tuning bandwidth shifts down in frequency by 80 MHz (the lower curve in Figure 5.21).

5.4 Semiconductor Device Grounding Methods

In the previous two sections we showed the importance of grounding semiconductor devices to realize optimum performance. In this section, we describe in detail methods for grounding devices, including the grounding of unpackaged and packaged MMICs.

5.4.1 Unpackaged MMIC Grounding

We use unpackaged MMICs to achieve the ultimate in performance or operation at high frequencies (usually above 18 GHz). For such applications, the degradation in performance caused by package parasitics is unacceptable. The typical MMIC is a thin piece of semiconductor a few thousandths of an inch (0.1 mm) thick, having active circuitry fabricated on its top surface. Ground can be referenced either to a MMIC's top or bottom conducting layer. The MMIC

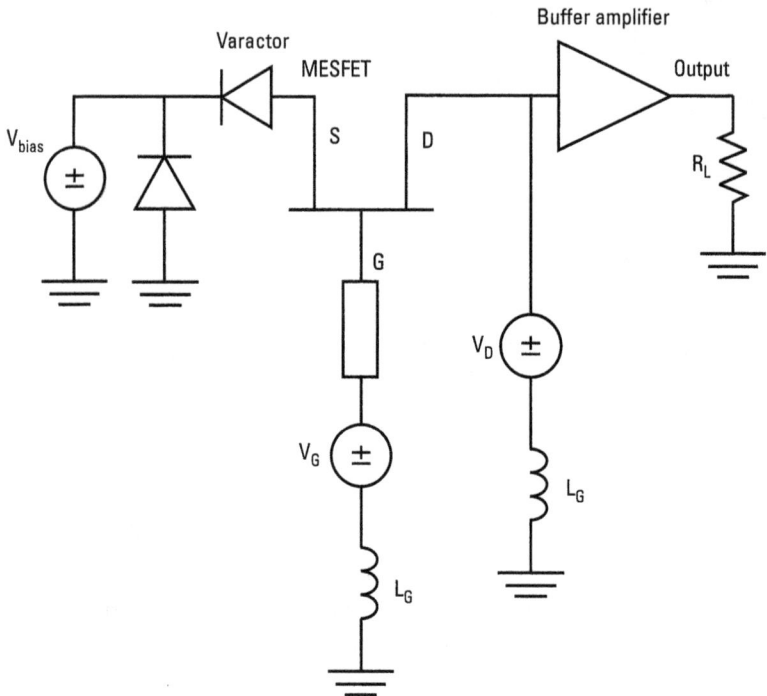

Figure 5.20 VCO with output buffer amplifier. V_{bias} tunes varactor and VCO frequency range. L_G is inductance in the ground path of the MESFET bias circuit. (*After:* [16].)

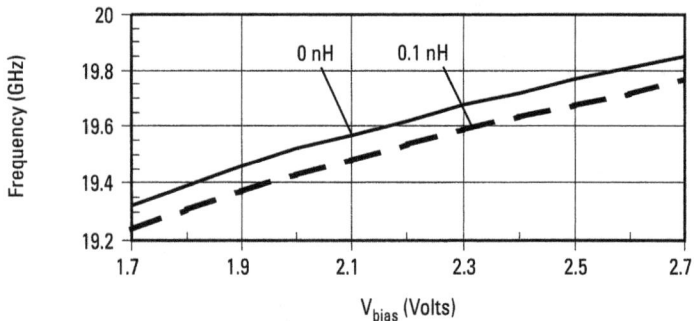

Figure 5.21 Buffered VCO output tuning range for $L_G = 0$ nH and 0.1 nH.

is attached to a conductive housing or circuit board ground plane using conductive epoxy or solder, and the housing or circuit board continues the ground path to the DC and RF ground(s) for the system. Properly grounding the active devices on the MMIC involves minimizing the path length along with inductance and resistance to ground.

For MESFET amplifiers, we know that minimizing source inductance to ground is critical—a topic discussed in some detail by DiLorenzo and Wisseman

[17]. Figure 5.22 shows three ways a MMIC can be grounded to a circuit board. An early approach, still implemented by a few MMIC designs, is to attach bond wires from the source contacts on the upper surface of the MMIC to the ground plane as shown in Figure 5.22(a). Three or four bond wires may be required to lower the source inductance sufficiently to stabilize a high gain amplifier. A more effective grounding method employs via holes through the MMIC, as illustrated in Figure 5.22(b). The via holes are formed as the MMIC is processed, and no wire bonds are required to ground the MMIC after it is attached to a housing or circuit board. Another approach that promises to have very low source inductance is *flip-chip mounting*, shown in Figure 5.22(c). The MMIC is flipped so its active side faces down towards the circuit board. The length of the ground path is very short even compared to via hole grounding.

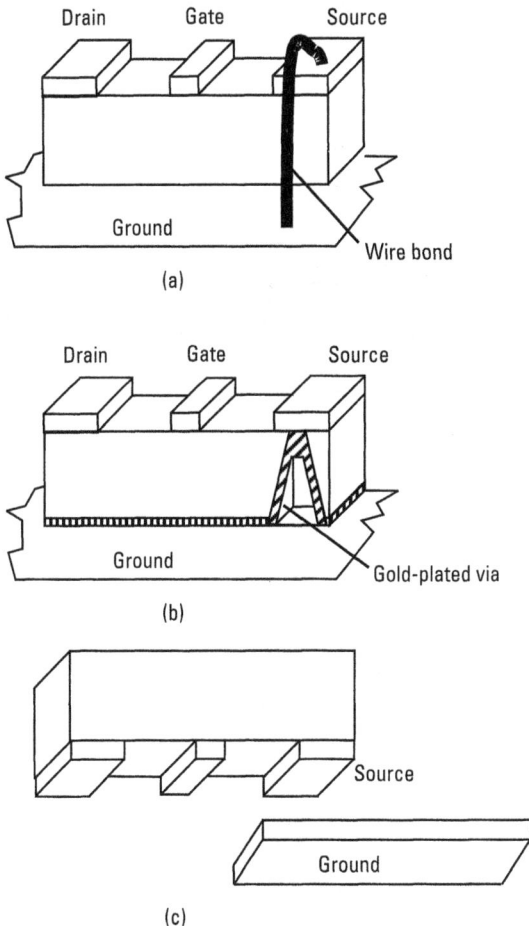

Figure 5.22 Source grounding of a MESFET integrated circuit: (a) wire bond; (b) conductive via; and (c) flip-chip mounting.

Figure 5.23 illustrates two methods for mounting MMICs with via hole grounding onto circuit boards. For relatively thin circuit boards, such as the thin-film, single-layer alumina or fused silica substrates that often are used at microwave and millimeter-wave frequencies, a pocket is routed in the circuit board as shown in Figure 5.23(a). The MMIC is attached with epoxy or soldered directly to the housing floor or circuit board ground plane. Very little additional inductance is added to the ground path using this grounding method. Direct attachment to a housing floor also provides excellent heat sinking. The DC and RF connections between the MMIC and circuit board transmission lines and bias conductors are made with wire bonds. Silverman discusses the advantages of this method further [18].

Figure 5.23(b) shows another method for attaching and grounding a MMIC to a circuit board. A metallized grounding pad is created on the top layer of the circuit board, the layer that includes the transmission line traces. An array of via holes is drilled beneath the pad, connecting it with the RF ground plane. The MMIC is epoxied to the pad. The quantity, spacing, and size of vias determine the inductance to ground and also the thermal resistance to the heat sink. Tischler and coworkers have examined via hole array grounding in some detail

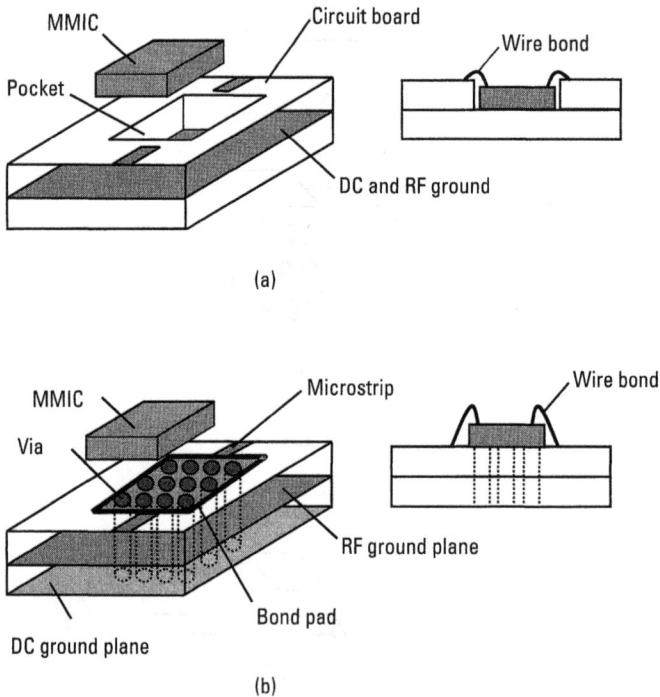

(a)

(b)

Figure 5.23 Grounding of a MMIC to a circuit board: (a) pocket placement allows direct mounting to RF ground plane or (b) vias through circuit board.

[19]. They used an electromagnetic simulator and found that above a certain critical frequency, determined by the spacing between the via holes, this grounding mechanism fails. They recommend that via pitch and diameter be chosen carefully to assure that the critical frequency is outside the passband of the circuit. Their results show that the pitch should be well below a half wavelength at the highest frequency of operation. They also suggest extending the edge of the metal pad no further beyond the outermost row of vias then required by the circuit layout rules.

5.4.2 Packaged MMIC Grounding

Packaged MMICs are used commonly for commercial and military low frequency (below 6 GHz) applications that demand low cost rather than the highest performance. Packaged MMICs are handled easily and can be attached to circuit boards along with resistors, inductors, and capacitors as part of a production solder reflow process. Unlike unpackaged devices, packaged MMICs do not require bond wire interconnections—the wires are inside the package. Figure 5.24 shows a conventional four-lead, pill-type package mounted to a circuit board. The RF input and gate DC bias voltage pass inside the package through a single lead. A second lead brings the drain DC bias to the MMIC, and allows the RF output to exit the package. The source leads are connected to circuit board traces that are grounded through via holes. In addition to the MMIC's source inductance, the package leads contribute inductance, as do the circuit board traces and vias.

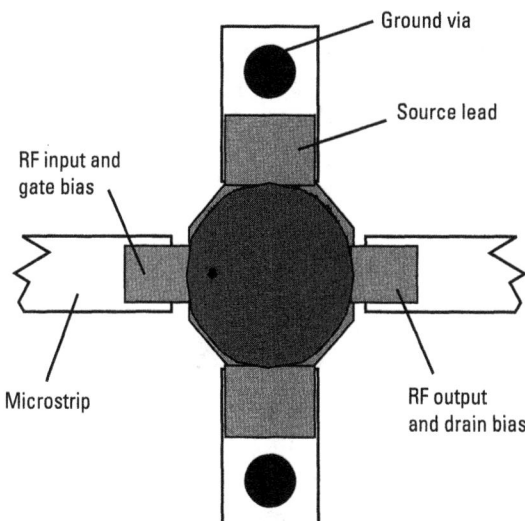

Figure 5.24 Packaged MESFET with source grounding.

A device's lead inductance can be large enough to dominate the total path inductance. For example, Wei et al. describe the behavior of an Agilent Technologies MGA 86576 amplifier on a 0.032-inch (0.81 mm) thick circuit board [14]. The S-parameters of the MMIC as provided by Agilent indicate the MMIC is unconditionally stable from 2 to 10 GHz. However, after the packaged MMIC is mounted on the circuit board, at least four vias are required to keep the device unconditionally stable over the same frequency range with the package lead inductance ignored. Once the leads of a package like that shown in Figure 5.24 are considered, the lead inductance dominates the ground path impedance. Even with eight via holes, the stability factor drops below unity from 2 to 7 GHz. A prudent circuit board designer would design and incorporate a circuit to stabilize this packaged device.

Figure 5.25 shows the interior of an eight-pin package. Two source leads are connected to the ground paddle in the center of the package onto which the MMIC is mounted. Bond wires connect the MMIC to the other six leads, which carry DC bias and RF signals. This package contributes inductance to the ground path through its leads and through the ground paddle. A simple equivalent circuit for the ground paddle is a parallel LC circuit to ground [20].

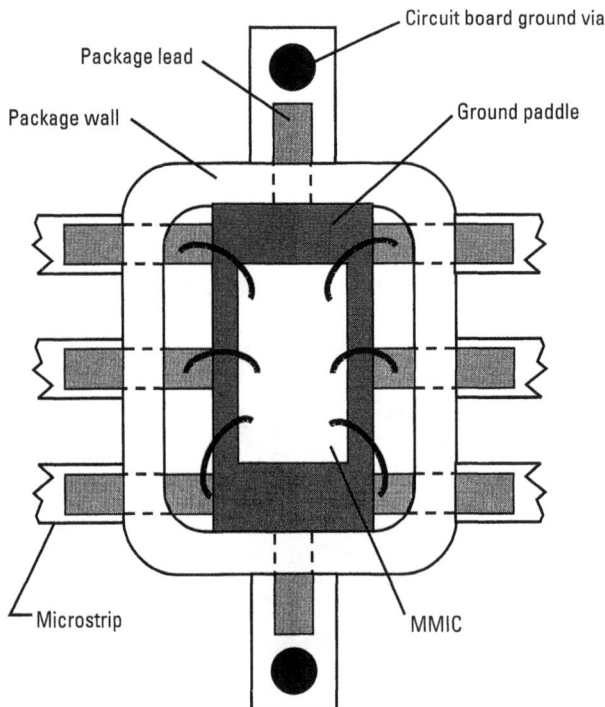

Figure 5.25 Eight-pin package.

5.5 Grounding of Microwave Subsystems

So far, we have focused our attention on individual active circuit components and the influence of grounding on their performance. In this section we investigate the grounding of microwave subsystems having more than one active component.

5.5.1 Grounding of Microwave Modules

When we need to ground more than one active microwave component, particularly modules with coaxial or waveguide interconnections, we almost always have to contend with multiple ground paths as shown in Figure 5.26(a). In this example, two active modules receive their DC bias from the same power supply. One wire runs from the positive terminal of the supply to the positive terminal of each module. Similarly, a single wire is the primary return from each

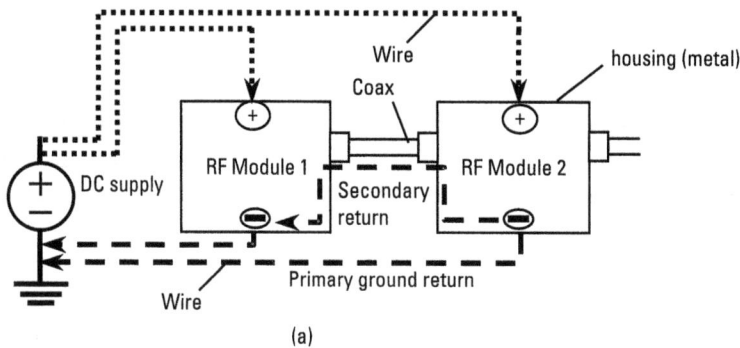

(a)

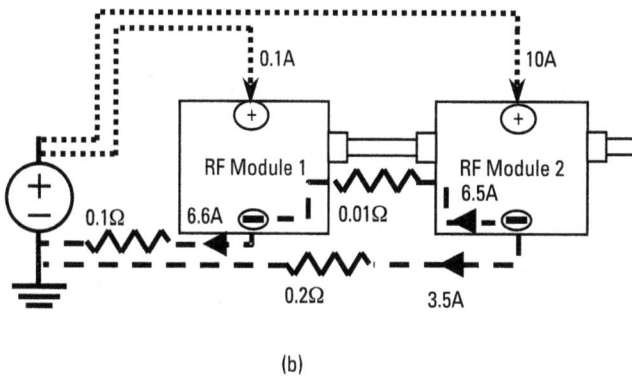

(b)

Figure 5.26 (a) Single point grounding of two modules is defeated by a coaxial cable interconnection. (b) Module 2 ground current flows along both return paths. (*After:* [21].)

module's negative terminal to the negative terminal of the power supply. If the modules were not interconnected by the coaxial cable, this would be a single-point grounding scheme, and the ground currents of the two modules would be completely isolated. However, as the figure reveals, the coaxial cable's ground conductor provides a secondary return path for the ground current of module 2, which passes through module 1. Since the impedances of module 2's primary and secondary ground paths are not zero, some of module 2's ground current will flow along both paths. The secondary path of module 2 includes module 1's primary return path, so a portion of module 2's ground current flows through a ground conductor that is common to both modules. In effect, the modules are multipoint grounded by their coaxial interconnection, and the ground currents of both modules are coupled.

Suppose module 1 is a low power amplifier that drives a power amplifier (module 2), as shown in Figure 5.26(b). Module 1 draws 0.1 amps of DC current, and module 2 draws 10 amps of current. We have assumed the ground wires from modules 1 and 2 have the same resistance per unit length, but that module 2's wire is twice as long. The coaxial cable, which has more conductor area, has much lower resistance than the wires. When we calculate the portion of module 2's ground current flowing through each path, we see that two-thirds of the current flows in the secondary path through the cable. This current causes the potential of the negative terminal of module 1 to rise to almost 0.7 volt, which may reduce its drain-to-source voltage bias and decrease its output power and gain (see Section 5.3.2).

A satisfactory solution to this problem is to connect the modules to a low impedance metal plate such as the floor of a housing as shown in Figure 5.27(a) [21]. A metal plate or housing floor of reasonable thickness will have a resistance that is a fraction of a milli-ohm. A metal plate has low inductance in addition to low resistance, so it makes a good DC, digital, and RF ground path. In Figure 5.27(b), it is apparent that nearly all the ground current from module 2 is now flowing through its primary ground path (the plate). One should still connect ground wires between the modules and power supply for safety, should the modules become disconnected from the plate while turned on.

5.5.2   Active Device Grounding in Mixed Signal Printed Circuit Boards

In Section 3.8, we discussed grounding of passive devices on microwave printed circuit boards. Grounding on circuit boards becomes even more important when active devices are involved, especially when RF, DC, and digital devices are used on the same circuit board. In general, as we described in Section 3.8, for a four-layer circuit board, it is best to use an RF ground plane (the second metal layer) to shield the currents flowing in the RF components from the power and digital currents flowing on the other layers [22]. While all surface mount

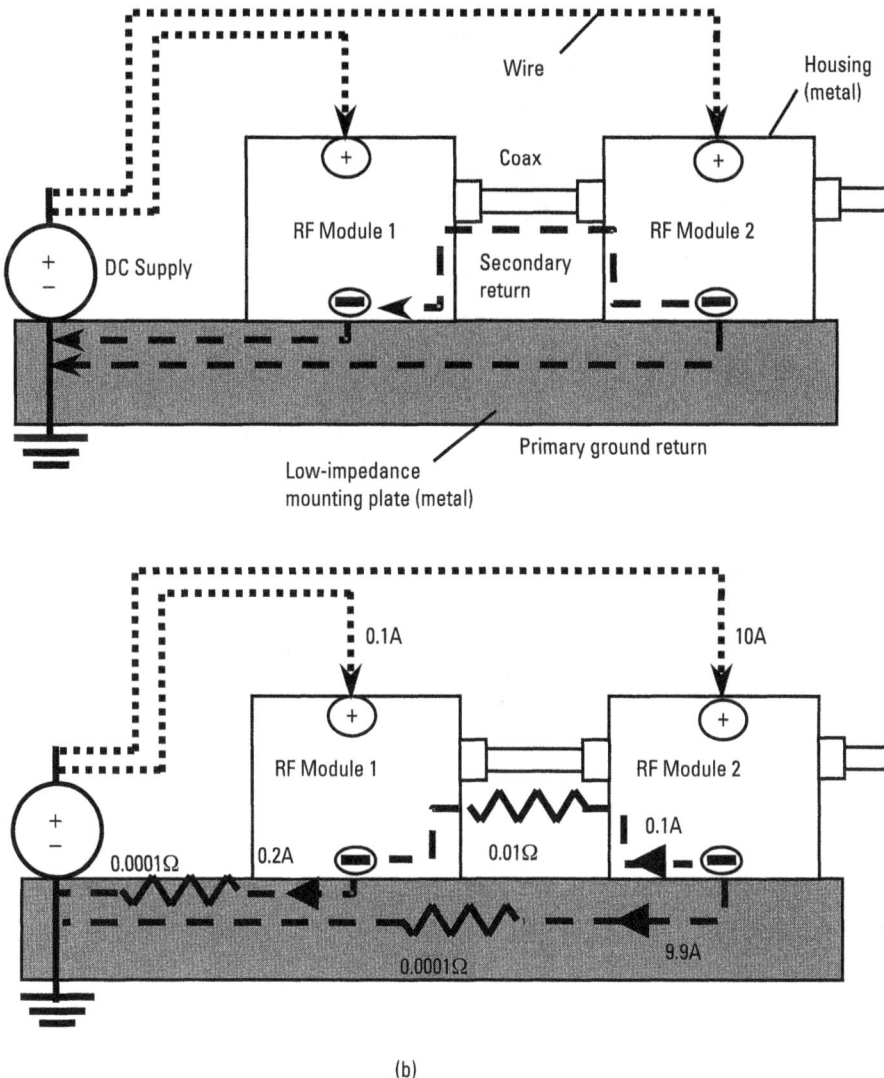

Wire

Housing (metal)

Coax

RF Module 1

+

DC Supply

Secondary return

RF Module 2

Low-impedance
mounting plate (metal)

Primary ground return

(b)

0.1A

10A

RF Module 1

RF Module 2

0.0001Ω

0.2A

0.01Ω

0.1A

0.0001Ω

9.9A

Figure 5.27 (a) A low impedance mounting plate makes a better primary ground current return path. (b) Very low resistance in primary return assures that little return current flows through coaxial conductor. (*After:* [21].)

components, whether RF, DC, or digital, are mounted on the top layer, one should route all ground paths to the DC/digital ground following the shortest path possible. To minimize inductance, multiple vias should be used for RF components. Surface mounted power amplifiers may require multiple vias to ground for heat dissipation. One should avoid having decoupling capacitors share ground vias, because the resulting common ground path will likely couple signals that were intended to be isolated.

Figure 5.28 illustrates a common problem that can develop in mixed signal circuit boards. In this example, two microwave amplifiers and their high current DC-to-DC converter are using a common ground path. The converter has a switching frequency measured in kilohertz. Its ground current is coupled to the source current of the two amplifiers through a common ground path that is likely to exist in a circuit board. Any impedance in the source path of each amplifier will develop a voltage that is modulated at the switching frequency of the DC converter. The bias modulation will cause a modulation to appear on the RF output of the amplifiers. For many microwave systems, this modulated output can degrade performance. When the amplifiers and converter are on the same circuit board, the coupling can be suppressed if the ground paths can be isolated. One method, shown in Figure 5.29, is to clear a moat around the ground plane beneath the converter and thus force the converter's ground current to take a return path that is removed from the amplifier chain.

In a related discussion, Scolio discusses basic switching-regulator layout techniques, with an emphasis on ways to keep the regulator's noisy power section from affecting the quieter devices on a circuit board [23]. He recommends creating two separate ground sections for the regulator, one for its power components and one for its quieter analog circuitry. He suggests maximizing the width and minimizing the length of the printed conductors comprising the power circuitry's ground path and also making the ground connections between the regulator's power circuitry on the top layer rather than using vias to drop

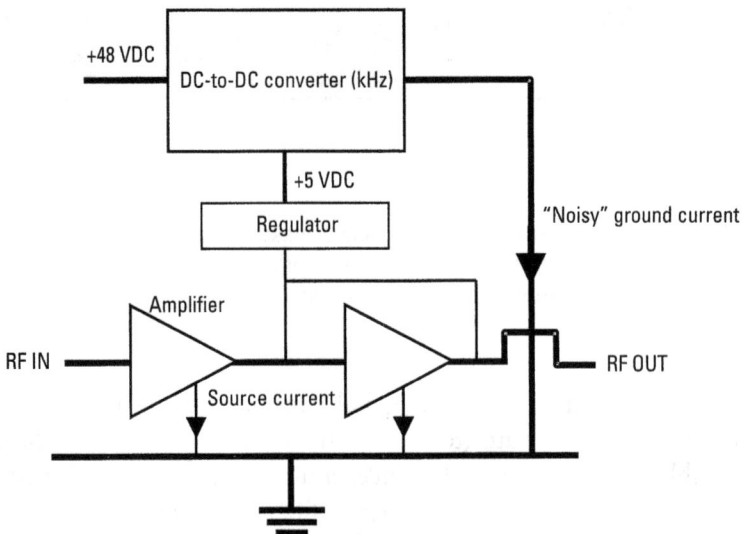

Figure 5.28 Multipoint grounding causes switching noise on DC-to-DC converter ground current to couple to RF amplifiers.

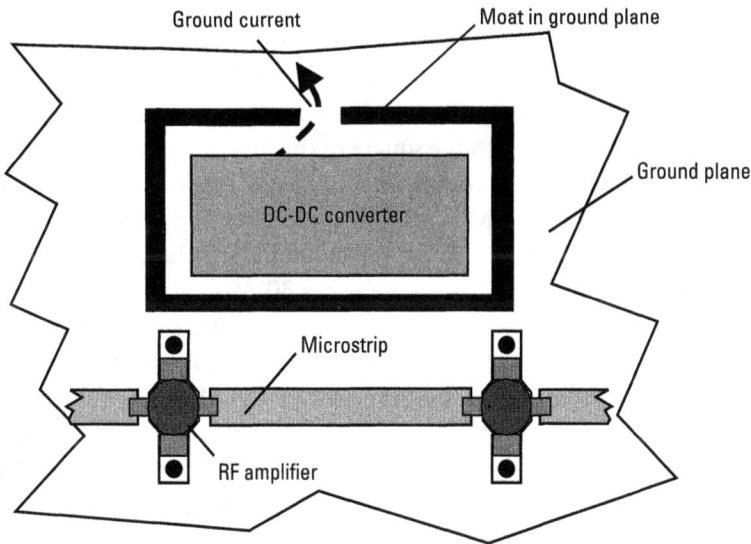

Figure 5.29 A moat in the ground plane around a DC-to-DC converter directs the converter's ground current away from other components.

down to an internal layer. Those connections should be confined to an isolated portion of the circuit board away from RF and other sensitive circuits.

As we discussed in Section 5.3.4, VCOs are particularly susceptible to poor grounding. VCO phase noise is sensitive to noise in its DC bias and tuning voltage supply lines, which can couple to the VCO's RF output through its ground path. When packaged VCOs are surface mounted to circuit boards, a number of precautions should be taken [24]. The microwave transistor's DC bias supply, the varactor's tuning voltage supply and the VCO's ground pins should be connected to the printed circuit board's DC ground plane. If the VCO package is pin-less, then it should be mounted to a metallized pad with multiple vias running to the circuit board ground plane [see Figure 5.23(b)]. In addition, high frequency chip decoupling capacitors should be inserted between the regulator bias voltage and ground. A phased locked loop, which incorporates a VCO in a feedback loop, requires care in grounding for the best performance. Bremer, Chavers, and Yu discuss the importance of grounding a surface mounted PLL in a Wi-Fi transceiver application [22].

5.5.3 Grounding of Amplifier Chains

We know from our discussion in Sections 5.3 and 5.4 that too much inductance in the source of a stable amplifier will make it oscillate. When amplifiers are cascaded as shown in Figure 5.30, even the best source grounding may not be adequate. Planar transmission lines such as microstrip, which often are used in microwave circuit boards, tend to radiate near circuit board discontinuities such as the leads of packaged amplifiers or the bond wire connections to MMICs. This radiated energy spreads out in all directions with intensity decreasing approximately in proportion to the square of the inverse of the distance from the discontinuity. Radiation can couple an amplifier's output to its input as indicated in Figure 5.30(a). A MMIC amplifier can have very high gain, and yet its microwave input and output may be separated by less then a tenth of an inch (2.5 mm). The amplifier in Figure 5.30(a) has 30 dB of power gain at some frequency (not necessarily in its operating band), and we assume that it remains unconditionally stable even inside a package when mounted on a circuit board. If we set the input drive level to the amplifier at –30 dBm, then the output will be 0 dBm, given that the amplifier is operating in its linear range. If the coupling between the package's output and input is –40 dB, the level of power fed back from the output present at the input is –40 dBm, which is 10 dB below the input drive level. Equivalently, we can say that the loop gain of the amplifier is 30 dB – 40 dB = –10 dB, which is less than 0 dB. Consequently, the coupled output level does not exceed that being supplied to the input, so this amplifier should remain stable.

In Figure 5.30(b), two of the 30-dB gain amplifiers have been cascaded, and now the loop gain of the pair is greater than 0 dB: 30 dB + 30 dB – 50 dB

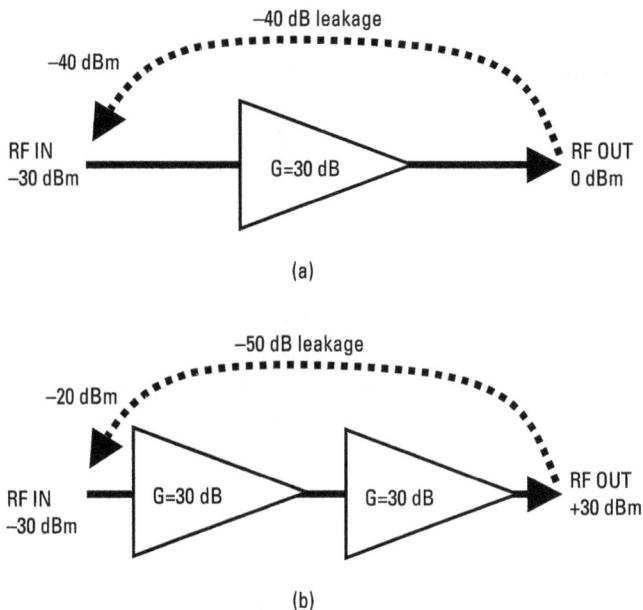

Figure 5.30 Unconditionally stable amplifier with reverse leakage path. (a) Stable operation: leakage from output is below input level. (b) Unstable operation: leakage from output exceeds input level.

= +10 dB. This pair of amplifiers will oscillate. We could reduce their loop gain below 0 dB by adding at least 10 dB of attenuation, which is not a desirable solution, especially if we need the forward gain. Alternatively, we could increase the isolation between the input of the first amplifier and the output of the second amplifier by increasing the separation between the amplifiers. But space on circuit boards is limited, and isolation increases approximately as the inverse of the distance squared. We will add only 6 dB of isolation if we double the amplifier's separation, and we need another 10 dB.

A common technique for stabilizing amplifier chains is to *channelize* the amplifiers as shown in Figure 5.31. Figure 5.31(a) shows two packaged amplifiers mounted on a circuit board. In Figure 5.31(b), we have covered the two amplifiers with a metal shield, which contacts the circuit board ground plane

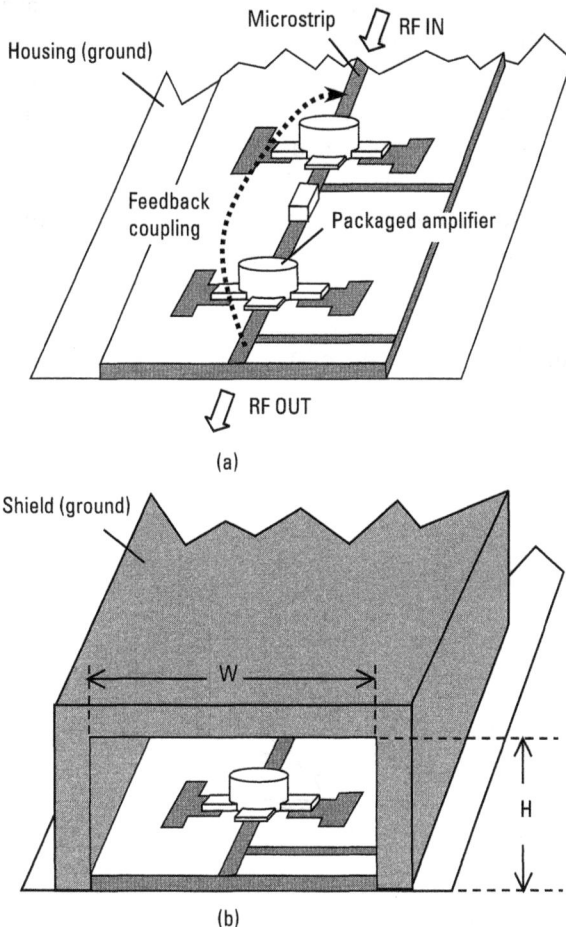

Figure 5.31 Packaged amplifier chain: (a) output-to-input feedback through radiative coupling; and (b) suppression of radiative feedback using a waveguide channel.

and forms a rectangular waveguide channel. Any radiated fields within the channel must take the form of waveguide modes. If we select the dimensions of the channel (w and H) such that it is cutoff to all waveguide modes, the isolation between the input and output of the amplifier chain increases greatly without any increase in its length. In effect, our placement of the shield over the circuit has grounded the unwanted radiation modes.

For a shield to be effective, it must start well before the input of the first amplifier and extend well past the output of the last amplifier. Otherwise, radiated energy can leak out the ends of the channel and pass around the outside of the shield. Additionally, the shield must make electrical contact with the ground plane of the circuit board to form a waveguide.

The approach shown in Figure 5.31(b) requires that slots be cut into the circuit board to enable the shield walls to touch the ground plane. However, for multilayer circuit boards with DC and digital signals often traveling in the layers beneath the RF signal layer, we may need to leave the circuit board intact. We can construct the channel as shown in Figure 5.32. The top and side walls of the shield are attached to metallized tracks on top of the substrate. Via holes connect the tracks to the ground plane. This channelization method provides less isolation than that in Figure 5.31(b), since gaps exist in the substrate between the grounding vias that form the narrow wall of the channel waveguide. These gaps are prone to radiate, especially if the transmission line is not straight (see Section 3.4).

With an electromagnetics circuit simulator, we can compute the isolation between the inside and outside of the channel as a function of via size,

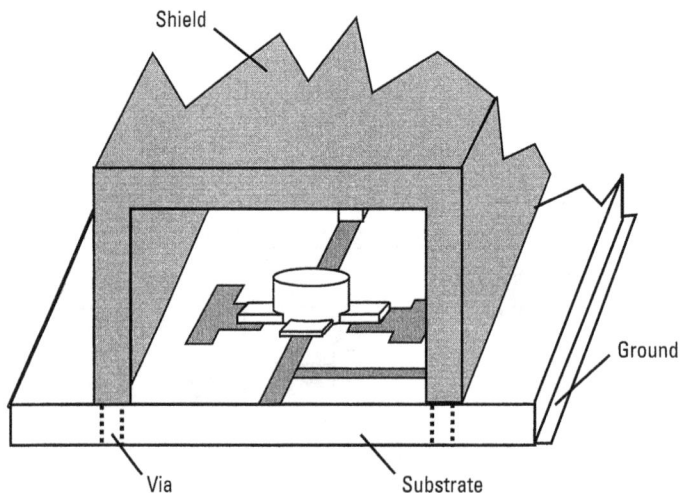

Figure 5.32 Mode suppression shield mounted on top of substrate and grounded with rows of vias.

separation, and the number of parallel rows of vias under each narrow wall. If we assume that a microstrip line connects the amplifiers, and if we further assume that the bulk of the electromagnetic energy resides in the substrate, we can predict isolation with the simplified parallel-plate waveguide models shown in Figure 5.33.

Figure 5.33(a) shows a portion of the substrate lying underneath one of the channel narrow walls and above the ground plane. To simplify the analysis, we have made the top and bottom surfaces of the substrate perfect electric conductors. We excite one side with a TEM mode, and determine the isolation by

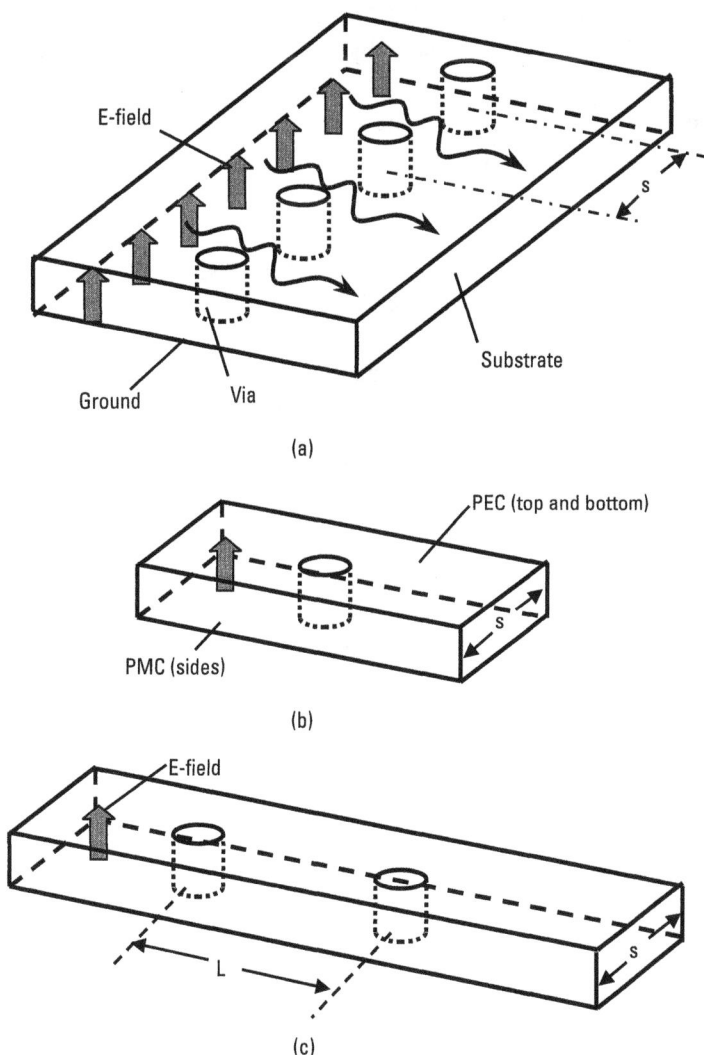

Figure 5.33 (a) Grounding vias in a circuit board. (b) Periodic cell for analysis. (c) Periodic cell for double row of grounding vias.

calculating the amount of energy that propagates through the vias to the other side. Since this is a periodic structure, we can model it as a single via in a parallel-plate waveguide, as shown in Figure 5.33(b). The sidewalls of this structure are perfect magnetic conductors (PMCs) in the simulation. Two rows of vias can be analyzed with the structure shown in Figure 5.33(c).

Figure 5.34 plots the isolation versus via spacing, s, in wavelengths for single and double rows of via holes. For one row of vias, the isolation barely exceeds 10 dB even for 20 vias per free-space wavelength. Two rows of vias improves the isolation, but because waves are reflected multiple times between rows, the reflected waves can combine such that the isolation becomes zero, as is the case for a row-to-row separation of $L = 1.2s$ and via-to-via spacing of $s = 0.17$ wavelengths. This phenomenon is avoided by drilling at least 10 vias per wavelength.

Exactly how the amount of via isolation impacts the coupling between the output and input of an amplifier chain depends on how much energy actually propagates from the output of the amplifier along the main transmission line towards the channel wall. For a microstrip transmission line, most of the energy resides within two line widths of the strip. Thus, at the edge of a channel, which typically is at least five line widths across, not much energy is present. Further, the energy must leak out of the channel at the output of the chain and then back in at the input, so it seems that 5- to 10-dB isolation may be adequate in many situations. However, if an amplifier chain continues to oscillate even after it has been carefully channelized, it is likely another leakage path exists, such as within or even underneath the circuit board.

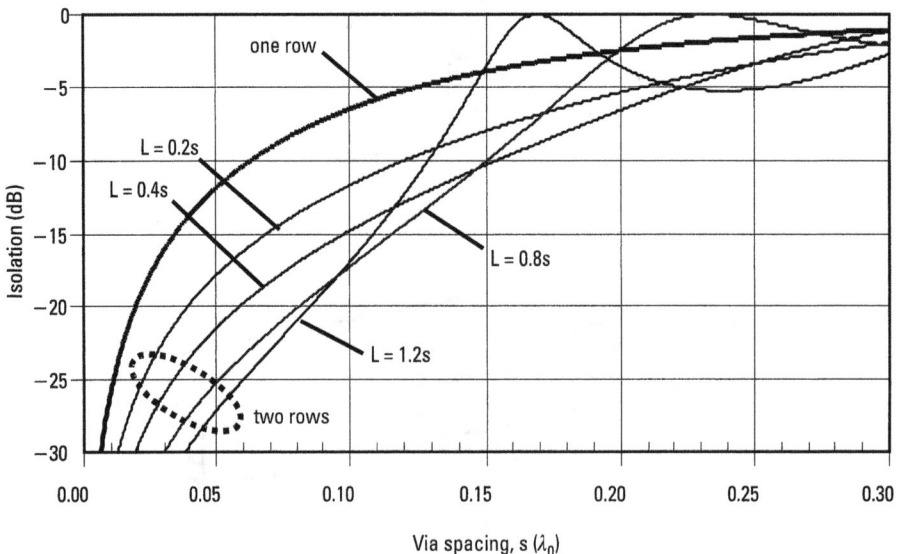

Figure 5.34 Substrate parallel-plate mode isolation from grounded vias. Via diameter: 0.08s.

5.5.4 Example: A Multilayer Circuit Board Transceiver

As a final example and summary of this chapter's grounding concepts, we study the multilayer circuit shown in Figure 5.35. This circuit performs C-band to K-band up- and downconversion and amplification of a digitally modulated signal. The C-band interfaces to the board are through vertically mounted coaxial transitions at the bottom left and right corners. The K-band connections are rectangular waveguide to microstrip transitions. On the far left side of the board, the transmit signal is filtered and upconverted before being amplified and outputted through the waveguide transition. On the far right, the received signal is amplified and downconverted. The local oscillator signals for the up- and downconverters are generated by two separate, multiplied VCO chains in the center of the board. Besides the RF electronics, the circuit board includes some digital circuitry to control the RF circuits and a DC-to-DC converter and regulator to bias the RF and digital circuits. Figure 5.36 shows the multilayer circuit board's construction.

The primary ground return is the floor of the metal housing. For a low impedance connection between the circuit board and the housing, conductive

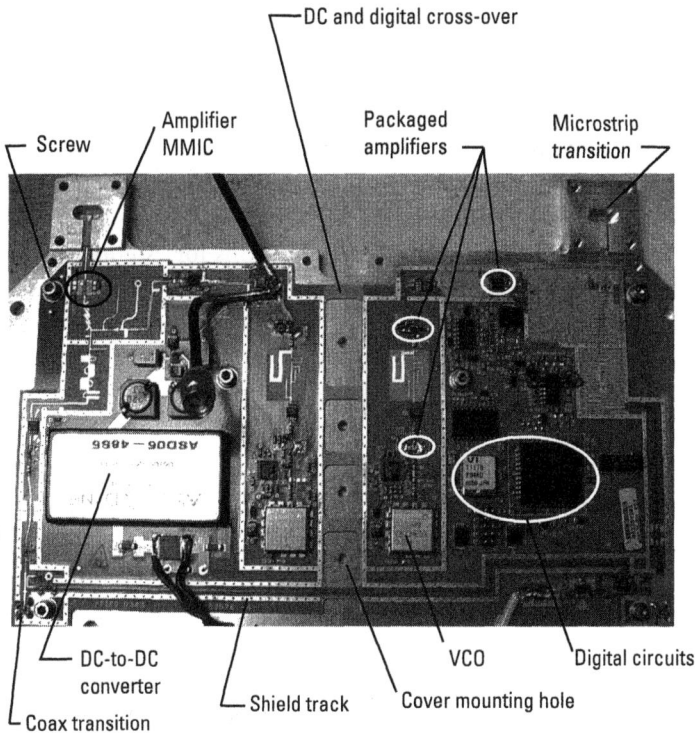

Figure 5.35 Multilayer circuit board with digital, DC power, RF, and millimeter-wave circuitry.

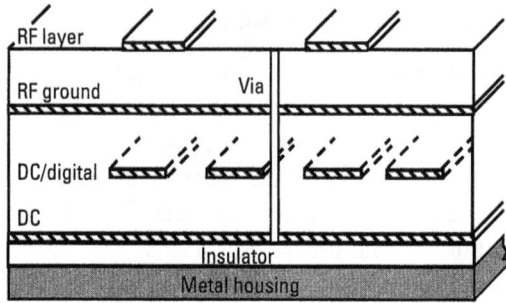

Figure 5.36 Cross-section of circuit board.

epoxy could be applied to the entire bottom surface of the circuit board, to bond it permanently to the housing. However, because the bottom layer must carry some DC signals besides ground, a thin insulator separates the circuit board from the housing floor. The ground connection between the circuit board and the housing floor is made with metal screws. Because the version of the circuit board shown in Figure 5.35 is missing screws near the VCOs, parallel-plate modes were able to propagate between the circuit board and metal floor causing coupling between the transmit and receive VCO circuits. Additional screws that were added in a later redesign of the circuit suppressed the parallel-plate modes.

Numerous shield tracks have been laid down on the top layer so that the C- and K-band circuitry can be well isolated with waveguide channels. In particular, the high power transmit and LO circuits must be isolated from the low power receive circuits. For additional isolation, the circuit board is divided in the middle, with only a few crossover connections for essential DC and digital signals. Shunt capacitors to ground are connected to the traces on both sides of the crossovers to suppress unwanted low frequency modes. Finally, a metal cover is screwed over the housing. A septum in the middle of the cover is screwed to the floor of the housing to separate the transmit and receive sides of the circuit board.

The components are attached to grounding pads on the top layer of the circuit board. Vias connect the pads to ground. The ground plane under the power supply includes an isolating moat with an opening towards the lower end of the circuit board. The K-band MMICs are unpackaged and epoxied directly to the top layer of the circuit board. Figure 5.37 shows details of the grounding pad for the transmit MMICs. The mounting area is densely packed with vias to ground for low inductance and thermal resistance. The waveguide probe, being on a multilayer circuit board, requires vias to suppress unwanted modes (see Section 4.3). In addition, a plastic insert in the waveguide channel forces the probe ground layer against the housing floor.

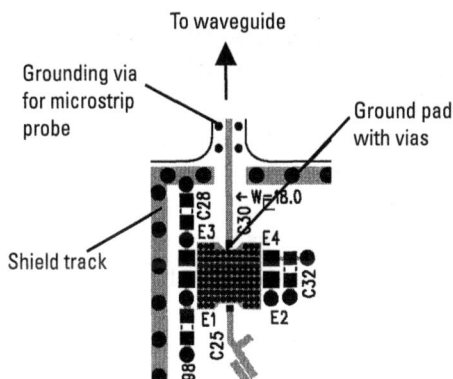

Figure 5.37 Detail of grounding pad for unpackaged MMICs.

References

[1] Holzman, E. L., and R. S. Robertson, *Solid-State Microwave Power Oscillator Design*, Norwood, MA: Artech House, 1992, Chapter 4.

[2] Maas, S. A., *Nonlinear Microwave and RF Circuits*, Norwood, MA: Artech House, 1988, Section 1.2.

[3] Maas, S. A., *Nonlinear Microwave and RF Circuits*, Norwood, MA: Artech House, 1988, p. 250.

[4] Holzman, E. L., and R. S. Robertson, *Solid-State Microwave Power Oscillator Design*, Norwood, MA: Artech House, 1992, p. 224.

[5] Bahl, I., and P. Bhartia, *Microwave Solid State Circuit Design*, 2nd ed., New York: John Wiley & Sons, 2003, Section 12.2.2.

[6] Bahl, I. and P. Bhartia, *Microwave Solid State Circuit Design*, 2nd ed., New York: John Wiley & Sons, 2003, p. 613.

[7] Eudyna Devices USA, Inc., http://www.us.eudyna.com/products/MWSelectionGuides/CBandInternalMatchedFETs.htm.

[8] Holzman, E. L., and R. S. Robertson, *Solid-State Microwave Power Oscillator Design*, Norwood, MA: Artech House, 1992, Section 2.6.2.

[9] Ha, T. T., *Solid-State Microwave Amplifier Design*, New York: Wiley-Interscience, 1981, p. 8.

[10] Ha, T. T., *Solid-State Microwave Amplifier Design*, New York: Wiley-Interscience, 1981, p. 37.

[11] Pengelly, R. S., *Microwave Field Effect Transistors: Theory, Design, and Applications*, Tucker, GA: Noble Publishing, 1994, Section 3.6.2.

[12] Ha, T. T., *Solid-State Microwave Amplifier Design*, New York: Wiley-Interscience, 1981, p. 67.

[13] Vendelin, G., "Feedback Effects on the Noise Performance of GaAs MESFETs," *International Microwave Symposium Digest*, March 1975, pp. 324–326.

[14]   Wei, Y. Y., P. Gale, and E. Korolkiewicz, "Effects of Grounding and Bias Circuit on the Performance of High Frequency Linear Amplifiers," *Microwave Journal*, Vol. 46, No. 2, February 2003, pp. 98–106.

[15]   Sirenza Microdevices, 0.1–12 GHz Low Noise pHEMT GaAs FET, SPF-2086T, http://www.sirenza.com/products_detail.asp?model=SPF-2086T&line=Low%20Noise%20Transistor&menu=amplifier.

[16]   Agilent Technologies, *Advanced Design System*, version 2004A.

[17]   Dilorenzo, J. V., and W. R. Wisseman, "GaAs Power MESFET's: Design, Fabrication and Performance," *IEEE Trans. on Microwave Theory and Techniques*, Vol. 27, No. 5, May 1979, pp. 367–378.

[18]   Silverman, L., "Mounting Technique Aids MMIC Performance," *Microwaves & RF*, Vol. 42, No. 8, August 2003, pp. 60–68.

[19]   Tischler, T., et al., "Via Arrays for Grounding in Multilayer Packaging—Frequency Limits and Design Rules," *2000 IEEE International Microwave Symposium Digest*, pp. 1147–1150.

[20]   Wartenberg, S., "RF Test Fixture Basics," *Microwave Journal*, Vol. 46, No. 6, June 2003, pp. 22–40.

[21]   Koh, C., "Grounding Schemes," October 30, 2001, unpublished memo.

[22]   Bremer, R., T. Chavers, and Z. Yu, "Power Supply and Ground Design for WiFi Transceiver," *RF Design*, November 2004, pp. 16–22.

[23]   Scolio, J., "Basic Switching-Regulator Layout Techniques," *EDN*, November 27, 2003, pp. 79–84.

[24]   Mini-Circuits, Inc., "Characterizing Phase Noise," *RF Design*, January 2003, pp. 58–59.

6

Antennas

An antenna differs fundamentally from the other devices in this book, because the currents flowing on its surface are meant to radiate and interact with other antennas. The source of these currents is often a transmission line, and thus, an antenna can be viewed as a load excited by current that flows across the antenna's conductors and returns to ground. Commonly, an antenna is designed assuming it exists in a region of free space, with no other conductive objects present. Yet, every antenna is attached to another conducting object such as the metal housing containing its transmitter or receiver, and most antennas sit near the surface of the Earth, which has significant conductivity. As currents flow on antennas, they are likely to interact with the currents that flow on such other conductors. Most of these conductors, whether connected or disconnected from the antenna behave as ground planes, and it is their influence on antenna performance that is the subject of this chapter.

6.1 Fundamental Concepts

An antenna is a structure that is designed to radiate electromagnetic energy efficiently. In other words, it is a conducting object on which electrons are accelerated so that they may transfer energy in the form of electromagnetic waves to free-space with as little loss as possible. As part of a microwave system, an antenna serves as the interface between the guided, nonradiating currents and fields on a transmission line and the plane waves in free space [1].

Figure 6.1 shows a dipole antenna and its connection to a source of RF current, J. Charges of opposite polarity accumulate at the ends of the dipole. Whatever the specific nature of the RF source, be it a digitally modulated

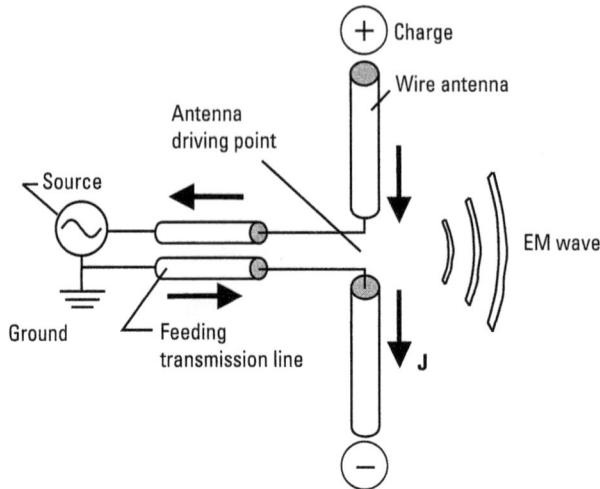

Figure 6.1 A transmission line delivers current **J** to an antenna, which dissipates energy through radiation.

communications transmitter circuit or a simple oscillator, it provides the electrons that radiate on the antenna surface. The interface between the source and antenna is a transmission line such as a coaxial line or microstrip. Electrons flow along the transmission line's signal conductor to the antenna, where they radiate, lose potential, and return via the transmission line's ground path to the source. The path the electrons take across the antenna may be quite complex, but we can view the antenna's driving point as a load impedance terminating the transmission line. The real part of the antenna impedance is called the *radiation resistance,* and it determines the level of radiated power in the far field of the antenna. The imaginary part of the antenna impedance accounts for stored energy in the antenna's near field. Ideally, we want to design an antenna for maximum radiation efficiency, which occurs when the imaginary part of its impedance is zero (resonance), and the real part is matched to the impedance of the transmission line.

Ground for an antenna can be defined as a reference potential and a source or sink of current. However, the antenna's radiative interaction with other structures complicates the situation. For example, the two arms of the half-wave dipole in Figure 6.1 are out of phase, with equal amounts of oscillating charge at each end. We arbitrarily can assign ground to the lower arm, which is connected to the source ground. The source ground and dipole ground generally are not at the same potential—current would not flow between them if they were. Now, if we place a metal sheet below the antenna, as in Figure 6.2, the antenna will interact with the metal sheet, inducing current to flow on the sheet. Although this current never flows to the antenna's source, its presence changes the antenna's radiation pattern and impedance, including the potential of the antenna ground.

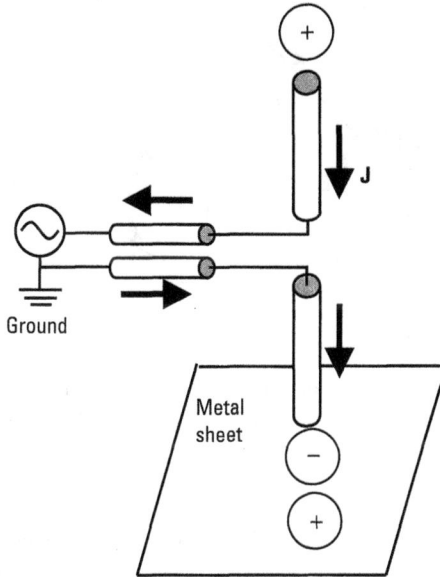

Figure 6.2 A ground plane perturbs an antenna's behavior. A negative charge at the end of the dipole induces a positive charge in the ground plane.

In effect, the metal sheet is part of the antenna, and we might even call it ground if it is connected to the source or antenna ground. Thus, compared to a nonradiating circuit, the behavior of an antenna is particularly sensitive to and may even depend on metal structures in its vicinity. However, it is important that the current flowing back to the source from the antenna terminals follow a low-loss, nonradiating ground path. Any impedance in this path will add to the antenna's impedance, changing the antenna's match to the transmission line and its ability to radiate efficiently.

There are two basic types of antennas: (1) *wire* antennas like the dipole shown in Figure 6.3(a), and *aperture* antennas like the horn shown in Figure 6.3(b) [2]. Obviously, the current flowing on the metal surface of either antenna is the source of radiation. The current distribution on the conductors of a wire antenna is relatively simple to deduce, and from it, we can determine the antenna's impedance and radiation pattern. In comparison, the radiating current distribution on a horn is complex; however, the electromagnetic field in the aperture can be approximated quite easily and used to calculate the horn's impedance and radiation pattern. Elliott [2] calls wire antennas *actual source antennas*, because we analyze the actual sources, the currents, flowing on the wires; he calls aperture antennas *equivalent source antennas*, because the electromagnetic field in the aperture is equivalent to the currents flowing on the aperture for impedance and pattern analysis.

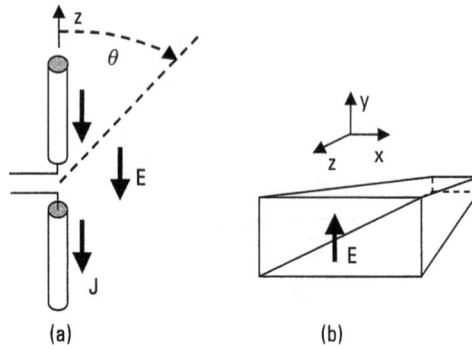

Figure 6.3 (a) Currents **J** on a wire antenna radiate. (b) Currents on a horn antenna aperture induce an aperture electric field **E**, an equivalent source of radiation.

If we measure the radiating characteristics of an antenna, we find that they depend on the source frequency, the observation angle (θ, ϕ) in space relative to the antenna and the polarization of the radiated electric field. The polarization is associated with the direction of the radiated electric field in space. If we consider a simple dipole comprising two oscillating charges like that shown in Figure 6.4, we see that its polarization is a function of the observer's location in space. The electric field is polarized vertically along the x-axis, but it rotates towards horizontal near $\theta = 0$ or π, depending on the distance from the midpoint between the two charges.

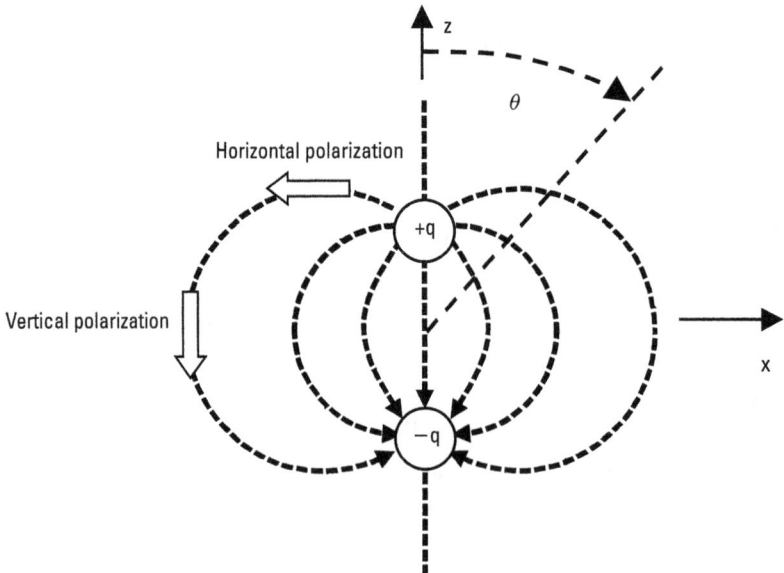

Figure 6.4 Electric field distribution for a two-charge dipole.

Since the charges at the ends of the wire dipole in Figure 6.3(a) do the radiating, its radiated field and polarization characteristics essentially are the same as that of the two-charge dipole. For the horn antenna in Figure 6.3(b), the polarization perpendicular to and in front of the aperture will be that of the aperture electric field, vertical in this case.

In general, antennas have two *principal planes*: (1) the plane of the antenna parallel to the electric field is the *E-plane*, and (2) the plane parallel to the magnetic field is the *H-plane*. The E-plane in Figure 6.3(a) is any plane parallel to and including the z-axis; the H-plane is the plane $\theta = \pi/2$. In Figure 6.3(b), the E-plane is the y-z plane, and the H-plane is the x-z plane.

We describe the spatial radiation characteristics of an antenna using a far-field radiation pattern (see Section 2.5). For example, Figure 6.5 shows

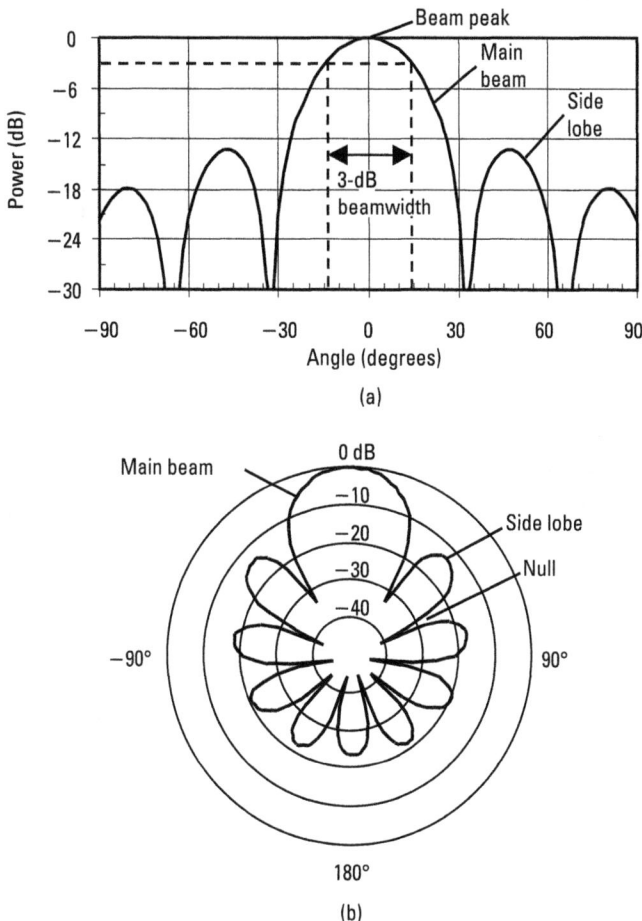

Figure 6.5 Antenna far-field radiation pattern plots: (a) Cartesian; and (b) polar.

Cartesian and polar plots of the radiation pattern for the same antenna. The Cartesian plot of Figure 6.5(a) gives a detailed picture of the radiation pattern in a restricted region of space, while the polar plot [in Figure 6.5(b)] is convenient for evaluating the overall pattern. Both of the patterns in Figure 6.5 are normalized such that the peak of the main beam is at 0 dB. Absolute directivity patterns are normalized to the directivity of an isotropic radiator, and the units of the *y*-axis are decibel-isotropic or dBi. The patterns in Figure 6.5 exhibit narrow *main beams*, where the directivity reaches a maximum, and numerous *side lobe* peaks straddling the maximum. The side lobes are separated by *nulls*, angles where there is no radiation at all. The *half power beamwidth* [see Figure 6.5(a)] is the angular width of the main beam, the angle between the two points on either side of the beam peak for which the directivity has dropped by half (3 dB). As a rule, peak directivity is inversely proportional to beamwidth, proportional to the size of the aperture (see Figure 6.6), and proportional to the square of the source frequency, as given by the formula [3]

$$D_{max} = 4\pi A_e / \lambda^2 \tag{6.1}$$

where A_e is the effective area of the antenna, which is related to the antenna's radiating surface area. *Aperture gain* (not to be confused with amplifier gain) is directivity (in decibels) minus any antenna losses (in decibels), including those from resistance and mismatch.

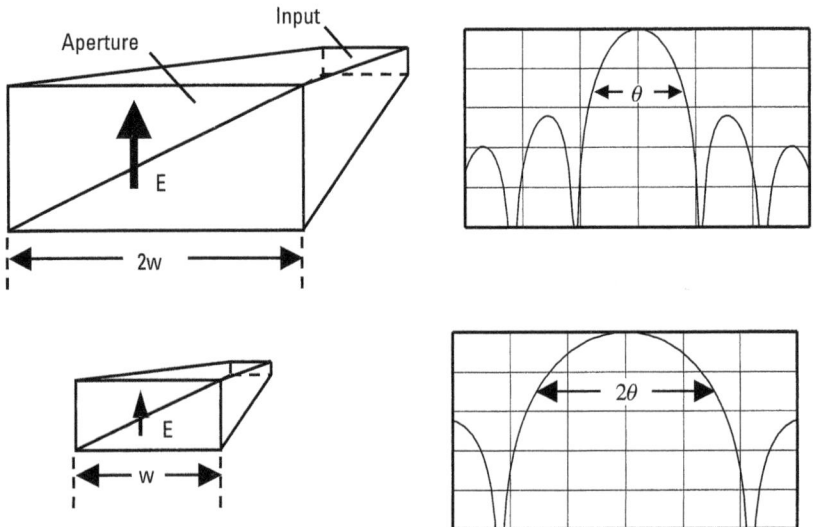

Figure 6.6 Antenna beamwidth is inversely proportional to aperture width: halving the aperture width (2w to w) doubles the H-plane beamwidth (θ to 2θ).

The response of a receiving antenna to an incident signal is measured by the amount of power flowing from its terminals. An antenna with low peak directivity, such as a dipole, tends to respond similarly to signals coming from many directions; while an antenna with high directivity, such as a large reflector, responds very strongly to a signal near the peak of its main beam but weakly to signals from elsewhere.

6.2 Interaction Between Ground Planes and Radiating Sources

Antennas commonly are designed to operate in empty space, even though they almost always end up radiating in the presence of other conducting structures, such as ground planes. Figure 6.7 shows an assortment of examples, including

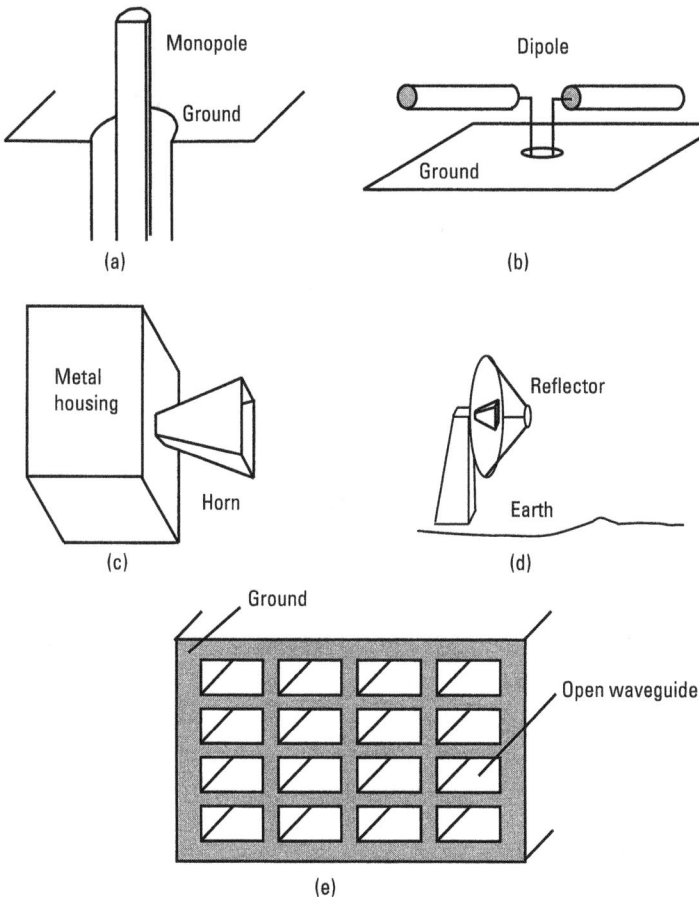

Figure 6.7 Antennas near ground: (a) monopole; (b) dipole; (c) horn; (d) reflector; and (e) open waveguide array.

monopole and dipole antennas over ground, a horn antenna that is part of a metal transceiver housing, a reflector antenna over the Earth, and an array of open rectangular waveguides cut into a metal faceplate. Each of these antenna's radiation and impedance characteristics are dependent to a varying degree on its location and distance from the nearby ground plane. Clearly, antennas rarely exist in isolation.

Two types of situations involving antennas and ground planes are of interest: (1) those for which the ground plane is an integral part of the antenna [Figure 6.7(a, c, e)], and (2) those for which the ground plane is not physically connected to the antenna [Figure 6.7(b, d)]. Every antenna, as part of a system, has at least one transmission line connected to its terminals. Besides providing a path for current to flow between the antenna and the rest of the system, the transmission line's structure may perturb the radiation characteristics of the antenna. For devices like cellular phones, the antenna is placed in close proximity to the cell phone transceiver electronics, which often are mounted on a circuit board having ground planes. The antenna designer must consider the interaction of the broad beamwidth cell phone antenna with the circuit board, which has unintentionally become part of the antenna. On the other hand, a reflector antenna is a high directivity antenna that relies on a precisely shaped and isolated ground plane (the reflector) to focus the radiation of a lower directivity feeding antenna such as a horn.

In this section, we discuss in general terms the effect of ground planes on radiating sources, including reflection by perfect and imperfect infinite ground planes, and diffraction by finite ground planes.

6.2.1 Perfectly Conducting, Infinite Ground Planes

A perfectly conducting ($\sigma = \infty$), infinite ground plane reflects an incident electromagnetic wave according to Snell's law, at an angle from the normal to the ground plane equal to the incident angle. Thus, if we place an antenna over a ground plane as in Figure 6.8, the *back wave*—the radiation from the antenna in the direction of the ground plane—will be reflected up into space. The reflected wave will add constructively or destructively with the direct wave, depending on the distance separating the antenna from the ground plane and the angle of observation. The higher is the antenna's directivity in the direction of the ground plane relative to that away from it, the more the reflected radiation will perturb the antenna's direct radiation. Conversely, an antenna that does not radiate a significant back wave will suffer little change to its radiation pattern when it is placed above a ground plane.

The Method of Images is a simple but effective technique we can use to compute the radiation pattern of an antenna over an infinite ground plane [4]. From Chapter 2, we know that the tangential electric field on a perfect

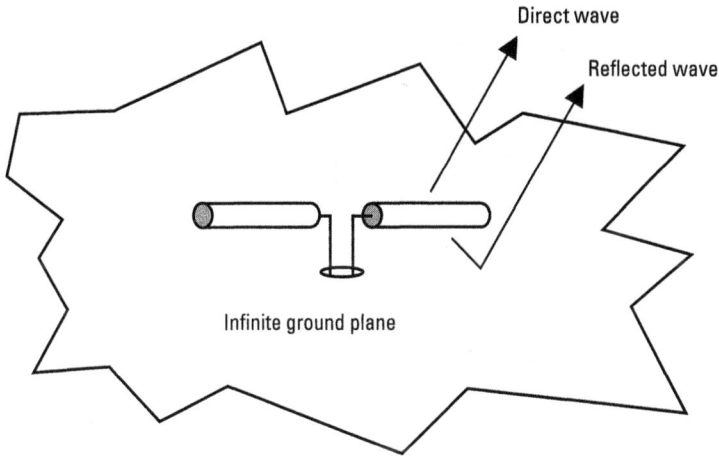

Figure 6.8 Antenna over an infinite, perfectly conducting ground plane. The ground plane reflects the antenna radiation.

conductor must be zero, and the normal electric flux density lines must terminate normally on the conductor's surface. Now consider the infinitesimal, vertically polarized electromagnetic source shown in Figure 6.9(a). The source is embedded in a perfect insulator with dielectric constant ε and is located a distance h above the surface of an infinite ground plane. If we observe the source in the far field at an angle θ, we will see two rays, a direct ray and a reflected ray, from the ground plane.

A pair of sources that radiates an equivalent field in the region above the ground plane ($z = 0$) is shown in Figure 6.9(b). We have removed the ground plane at $z = 0$ and placed an artificial source, the real source's image, at $z = -h$. The placement and orientation of the image are chosen to satisfy the boundary condition on the ground plane at $z = 0$. In other words, if the radiation from the vertical source and its image yields a continuous normal field at $z = 0$, then from our observation point at an angle θ, it will appear that the image's ray is identical in amplitude and phase to the ray reflected by the ground plane in Figure 6.9(a).

Figure 6.10 shows an infinitesimal horizontally polarized source above a ground plane. In this case, the tangential electric field must be zero on the surface of the ground plane, which requires that the image be identical to but directed opposite to the source.

We can write an equation for the radiated far field of each pair of sources by summing the contributions from the source and its image. Figure 6.11 shows the relationship for the vertical source and its image as seen by an observer at a distance far greater than $2h$. Only a theta component of the far field exists, since the phi direction is orthogonal to the source.

Source

z = h

$\sigma = 0, \varepsilon$

z = 0

Ground $\sigma = \infty$

(a)

Source

z = h

$\sigma = 0, \varepsilon$

z = 0

$\sigma = 0, \varepsilon$

z = −h

image

(b)

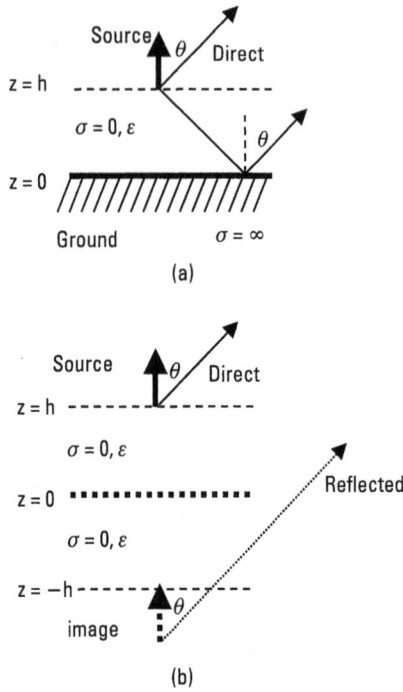

Figure 6.9 Vertical electric source (a) over a ground plane, and (b) the equivalent problem for $z = 0$ with an image replacing the ground plane.

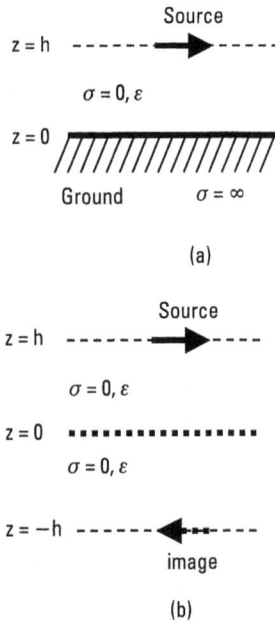

Source

z = h

$\sigma = 0, \varepsilon$

z = 0

Ground $\sigma = \infty$

(a)

Source

z = h

$\sigma = 0, \varepsilon$

z = 0

$\sigma = 0, \varepsilon$

z = −h

image

(b)

Figure 6.10 Horizontal electric source (a) over a ground plane, and (b) the equivalent problem for $z = 0$ with an image replacing the ground plane.

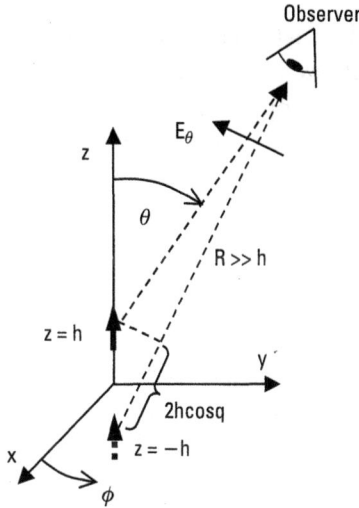

Figure 6.11 Two-element array and observer in the far field.

$$E_{\theta V} = s_\theta(r,\theta,\phi)\left(e^{jkh\cos\theta} + e^{-jkh\cos\theta}\right) = 2s_\theta(r,\theta,\phi)\cos(kh\cos\theta) \quad (6.2)$$

where $k = 2\pi/\lambda$, $s_\theta(r,\theta,\phi)$ is the source's theta component of electric field, and the bracketed term is the *array factor* that arises from the spatial separation factor, $2h\cos\theta$, between the source and image. In effect, the source plus its image radiate as a single element with the far-field radiation's apparent point of origin at $z = 0$, the *phase center*. In summary, for a vertically polarized source above a perfectly conducting ground plane, we can use (6.2) to determine the theta component of the radiated far-field pattern above the ground plane given the source's far-field pattern without the ground plane present.

For the horizontal source in Figure 6.10, we have both theta and phi components of the electric field far from the source:

$$E_{\theta H} = s_\theta(r,\theta,\phi)\left(e^{jkh\cos\theta} - e^{-jkh\cos\theta}\right) = j2s_\theta(r,\theta,\phi)\sin(kh\cos\theta) \quad (6.3a)$$

$$E_{\phi H} = s_\phi(r,\theta,\phi)\left(e^{jkh\cos\theta} - e^{-jkh\cos\theta}\right) = j2s_\phi(r,\theta,\phi)\sin(kh\cos\theta) \quad (6.3b)$$

In this case, the source and image contributions subtract, so that as h approaches zero, the source and image field cancel on the surface of the ground plane.

We have derived (6.2), (6.3a), and (6.3b) assuming the source is infinitesimal in size. For finite sized sources, these equations still hold provided we define h as the distance to the source's phase center without the ground plane present.

Figure 6.12 plots the far-field theta component of the electric field for vertical and horizontally polarized isotropic antennas at different distances h above a ground plane. Since an isotropic radiator has a nondirective pattern, it is clear from the plots in Figure 6.12 that the ground plane changes the radiation

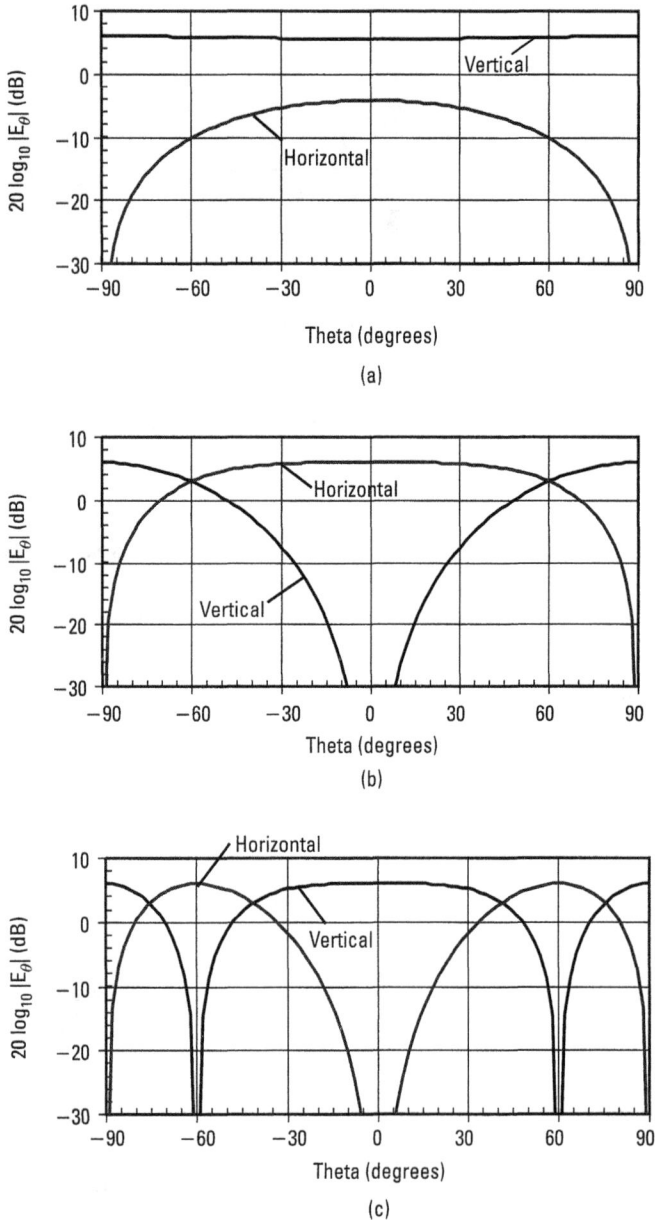

Figure 6.12 Far-field patterns of vertical and horizontal isotropic antennas above a ground plane: (a) $h = 0.05\,\lambda$; (b) $h = 0.25\,\lambda$; and (c) $h = 0.5\,\lambda$.

pattern dramatically, making it 6 dB more directive at some angles and causing nulls to appear at other angles. The locations of the nulls vary with frequency. In Figure 6.12(a), the antennas are just 5% of a wavelength above the ground plane. The vertical antenna pattern is reinforced by 6 dB, and the horizontal antenna is nearly shorted out. When we increase the height to a quarter wavelength [see Figure 6.12(b)], the situation is somewhat reversed, with the vertical antenna having a null at zenith (along the *z*-axis), and the horizontal antenna having 6-dBi directivity. The vertical and horizontal antennas exchange extrema at zenith every quarter wavelength, but more and more pattern nulls start to appear as the height above the ground plane increases. When real antennas are substituted for the isotropic antennas, this interfering effect of the ground plane remains.

We can make use of the just completed discussion to explain *multipath interference*, a phenomenon that occurs in microwave communications and radar systems that rely on wave propagation parallel to and near the surface of an imperfect ground (the Earth). Figure 6.13(a) shows a fixed communications link in which a transmitting antenna radiates in the direction of a receiving antenna. We approximate the Earth's surface by a perfect conductor to simplify the analysis (see Section 6.2.2 for a discussion on imperfect ground planes). The primary or direct signal travels straight from the transmitting antenna to the receiving antenna. However, the transmitting antenna radiates in all directions, and a secondary signal is incident on the Earth midway between the two antennas at an angle that depends on the antenna's height above the ground. It is reflected forward in the direction of the receiving antenna. Since the reflected wave travels a longer distance than the direct wave, the two waves are received at different times and generally have different phases. At a particular frequency the reflected wave will be 180° out of phase with the direct wave, and the two waves will interfere destructively. The resulting decrease in signal strength at the receiver may be enough to cause a link dropout (i.e., insufficient signal-to-noise ratio at the receiver for the communications link to function).

Figure 6.13(b) shows the equivalent problem using image theory, assuming a vertical polarized antenna. Rather than looking at direct and reflected paths, we consider only the direct path between the phase centers of the transmitting antenna/image and receiving antenna/image pairs. The distorted antenna pattern of the source and image array accounts for the effect of the ground plane per (6.2). Obviously, the correct placement of the antennas above the ground is critical for achieving optimum link performance: we do not want a null in the transmit pattern in the direction of the receiving antenna and vice versa. However, if the link bandwidth is large, it may be difficult to locate the antennas such that dropouts are avoided at all operating frequencies. One way to solve the multipath problem is to use an array of antennas for transmitting and receiving. The location of each element of the array is different, and so is its

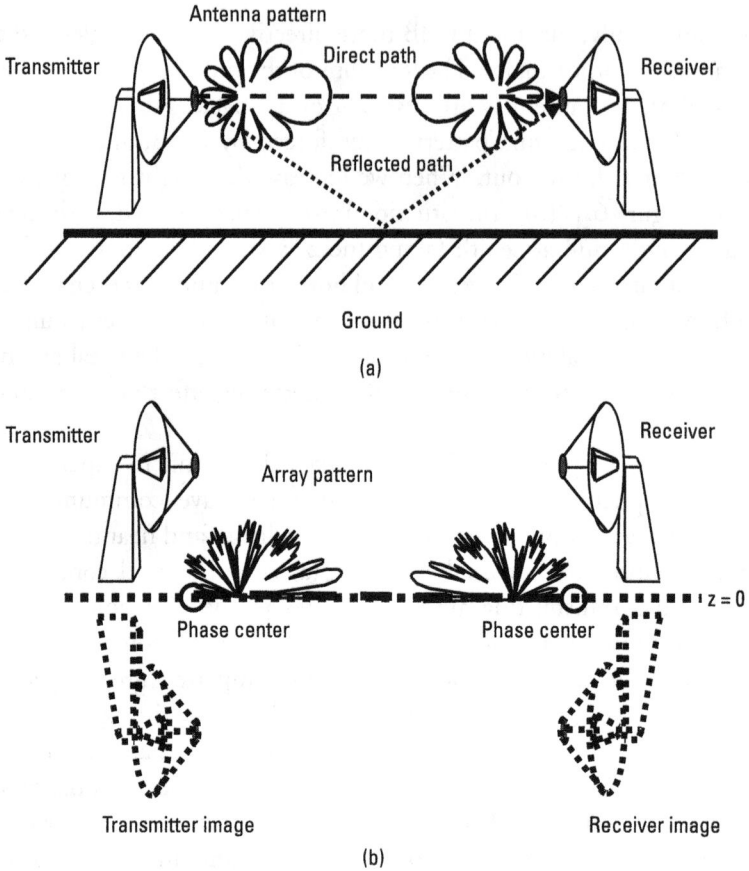

Figure 6.13 Multipath interference arises when (a) secondary ground reflection destructively interferes with the direct radiation at the receiving antenna. (b) An equivalent scenario using image theory.

multipath response. Consequently, the system can adapt at each frequency by selecting the pair of transmitting and receiving array elements giving the best link performance.

6.2.2   Imperfect Ground Planes

The high conductivity of metals like copper and aluminum means they behave essentially as perfect conductors with infinite conductivity at frequencies up into the millimeter-wave region. However, antennas are frequently located above imperfect ground planes such as the Earth, which have finite conductivity. Although we may still use image theory to compute the far-field pattern of an antenna above an imperfect ground, we cannot merely add or subtract the source and an identical image term to compute the far field. Unlike a perfect ground plane, which reflects all of an incident wave, an imperfect ground plane will absorb a portion of the electromagnetic energy incident upon it. From the standpoint of image theory, this behavior causes the image radiation to be attenuated in amplitude and shifted in phase relative to that in a perfect ground plane.

Figure 6.14 illustrates the situation for a vertically polarized source above an imperfect ground plane. In Figure 6.14(a), we see that a portion of the incident ray is transmitted into the ground plane, which causes the amplitude of the reflected ray to be less than that of the incident ray. The reflection coefficient for the reflected ray is given by [5]

$$\Gamma_V = \frac{\varepsilon_c \cos\theta - \sqrt{\varepsilon_c - \sin^2\theta}}{\varepsilon_c \cos\theta + \sqrt{\varepsilon_c - \sin^2\theta}} \tag{6.4}$$

where $\varepsilon_c = \varepsilon_0(\varepsilon - j\sigma/\omega\varepsilon_0)$, and we have assumed that $\mu = \mu_0$. We can write the radiated field of the vertical source over an imperfect ground by modifying (6.2):

$$E_{\theta V} = s_\theta(r,\theta,\phi)\left(e^{jkh\cos\theta} + \Gamma_V e^{-jkh\cos\theta}\right) \tag{6.5}$$

For a perfect conductor, $\sigma = \infty$, $\varepsilon_c = -j\sigma/\omega$, $\Gamma_V = 1$, and (6.5) reduces to (6.2).

Similarly, the reflection coefficient for a ray incident from a horizontal source on an imperfect ground plane is [5]

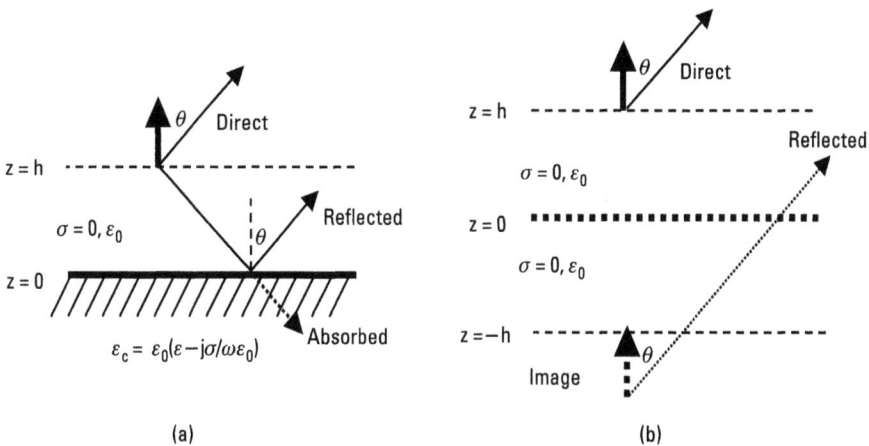

Figure 6.14 Electromagnetic ray emanating from a vertical electric source (a) over an imperfect ground plane, and (b) the equivalent problem for $z \geq 0$ with an image replacing the ground plane.

$$\Gamma_H = \frac{\cos\theta - \sqrt{\varepsilon_c - \sin^2\theta}}{\cos\theta + \sqrt{\varepsilon_c - \sin^2\theta}} \tag{6.6}$$

and (6.3a) and (6.3b) for the radiated field become, more generally

$$E_{\theta H} = s_\theta(r,\theta,\phi)\left(e^{jkh\cos\theta} + \Gamma_H e^{-jkh\cos\theta}\right) \tag{6.7a}$$

$$E_{\varphi H} = s_\varphi(r,\theta,\phi)\left(e^{jkh\cos\theta} + \Gamma_H e^{-jkh\cos\theta}\right) \tag{6.7b}$$

These equations reduce to (6.3a) and (6.3b) when $\sigma = \infty$. If we compare the electric field equations for the imperfect ground with those for perfect ground, we see that while it is possible to achieve total cancellation or a doubling of the field at certain angles with perfect ground, the presence of the reflection coefficient in the expressions for the field over imperfect ground means there will be a nonzero minimum, not a null, in field strength and a maximum that is less than double that of the source alone. Further, since Γ_V and Γ_H are in general complex valued, there will be a phase shift applied to the image terms, and the peaks and nulls in the patterns will occur at observation angles for the imperfect ground that are different from those of the perfect one [5]. Lastly, because the vertically and horizontally polarized wave reflection coefficients are different for the same angle θ, imperfect ground planes also can depolarize waves that combine both polarizations such as a circularly polarized incident wave, as described by Stutzman [6].

6.2.3   Finite Ground Planes

An infinite ground plane is an idealization, as all ground planes are finite in extent. However, if a ground plane extends sufficiently beyond the source, at least several times the length of the source, and if the source is not too far above the ground plane, the infinite ground plane assumption is a reasonably good approximation [7].

Whether finite or infinite in extent, all ground planes reflect electromagnetic waves, but the edges of a finite ground plane also *diffract* incident waves. As shown in Figure 6.15, *diffraction* is a phenomenon that enables electromagnetic waves to follow a path around an obstacle [8]. Currents flowing to the edge of a finite ground plane will excite radiation radially outward from the edge in many directions. In effect, the edge radiates like a line source. There is an extensive body of literature on diffraction and the geometric theory of diffraction (GTD), which is used to predict the precise behavior of diffracted fields. This theory enables one to calculate the diffraction coefficient of an edge much like

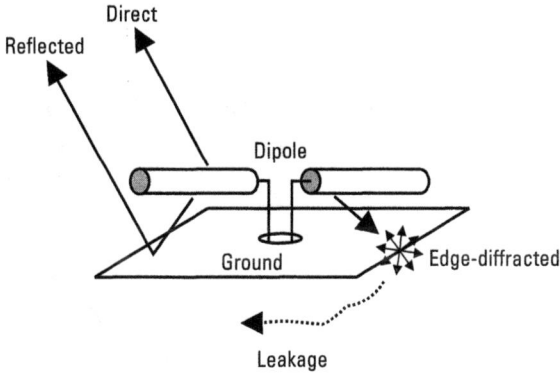

Figure 6.15 Finite ground plane edge diffraction of radiated wave causes radiation radially outward from edge, including leakage behind the ground plane.

one calculates the reflection coefficient at an interface between two different materials. A detailed discourse on the principles of GTD is beyond the scope of this book, but interested readers are encouraged to study Balanis' thorough treatment [9]. We examine the effects of diffracting ground planes on specific antennas in Sections 6.3 and 6.4.

6.3 Wire Antennas over Ground Planes

In this section, our discussion narrows as we study the behavior of monopole and dipole wire antennas over ground planes.

6.3.1 Fundamentals

Wire antennas (other than arrays) tend to have broad beamed, low directivity radiation patterns. The wire dipole, shown in Figure 6.16, has a torus-like, omnidirectional far-field radiation pattern. It has nulls on axis, and at zenith and nadir. Because of these characteristics, we would expect the pattern of a dipole oriented parallel to a ground plane to be perturbed more than that of one oriented perpendicular to the ground plane.

A dipole radiates most efficiently when it is about half a free-space wavelength long. A half-wavelength dipole's peak directivity is 1.64 (2.15 dBi), and its radiation resistance is about 73 ohms [10].

A monopole antenna is one-half of a dipole, and monopoles are almost always mounted above ground planes, as shown in Figure 6.17. The radiation pattern of a quarter-wave monopole above an infinite ground plane is identical to that of a dipole in free-space. Figure 6.17(c) plots the E-plane pattern, which when rotated about the z-axis has the shape of a toroid sliced in half. Because the

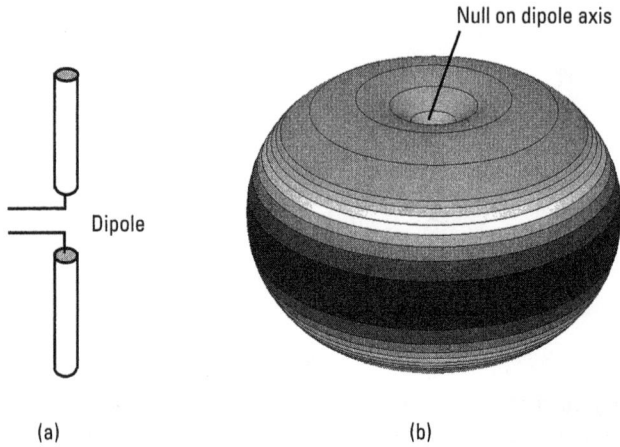

Figure 6.16 (a) A dipole antenna produces (b) an omnidirectional radiation pattern with a null on the dipole axis.

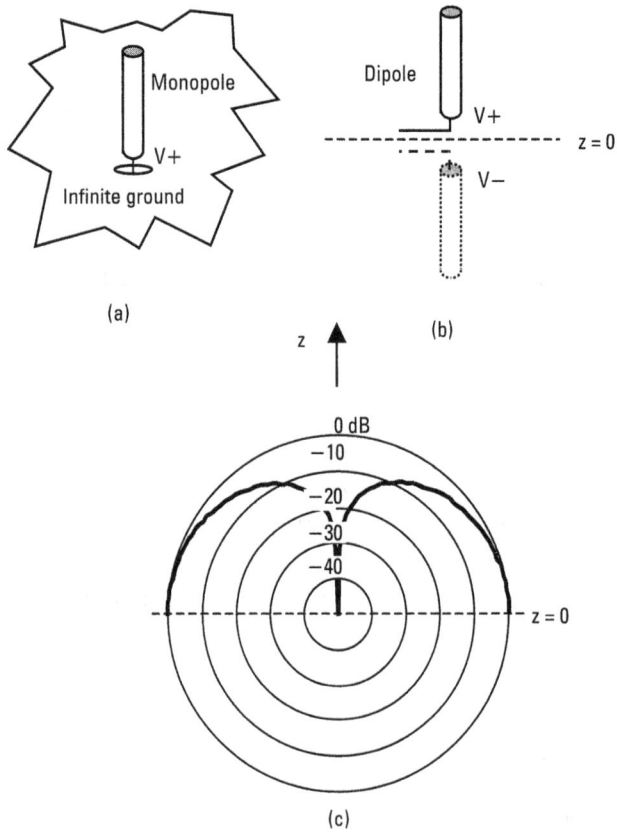

Figure 6.17 (a) Monopole over infinite ground (b) modeled as a dipole using image theory. (c) E-plane radiation pattern.

ground plane prevents all radiation below $z = 0$, the peak directivity is twice that of a dipole in free-space.

6.3.2 Monopoles and Dipoles over Ground

The monopole over a ground plane is one example of a ground-plane antenna, in which the ground plane is a necessary part of the antenna [11]. Weiner et al. have written a book on the thin, quarter-wave monopole over a circular ground shown in Figure 6.18(a) [12]. They focus on computing the monopole radiation pattern and input impedance as a function of the ground plane's radius. Figure 6.18(b) plots the radiation pattern of the monopole for three different values of radius r. For the ground plane extending to infinity, the pattern essentially is the same as the pattern in Figure 6.17(c), with a maximum directivity of 3.28 (5.15 dBi) at $\theta = 90°$. There is no ground plane for $r = 0$, and the maximum remains at $\theta = 90°$; however, without the imaging effect of the ground plane, it is reduced by roughly half to 1.5 (1.76 dBi). When the radius is finite, the diffracting edges interact with the primary field and cause the peak in directivity to shift towards 0°. For $r = 2\lambda/\pi$ in Figure 6.18(b), the peak angle is 40°, with a directivity of 2.46 (3.91 dBi).

In Figure 6.19, we see that as the ground plane radius increases, the directivity, beam pointing angle, and input resistance fluctuate, eventually converging to the infinite radius values. Consequently, the ground plane radius must be chosen carefully to optimize performance for this antenna, especially if a small radius is desired.

Another example of a ground plane antenna is the monopole over a conical ground plane shown in Figure 6.20(a). Figure 6.20(b) plots the radiation

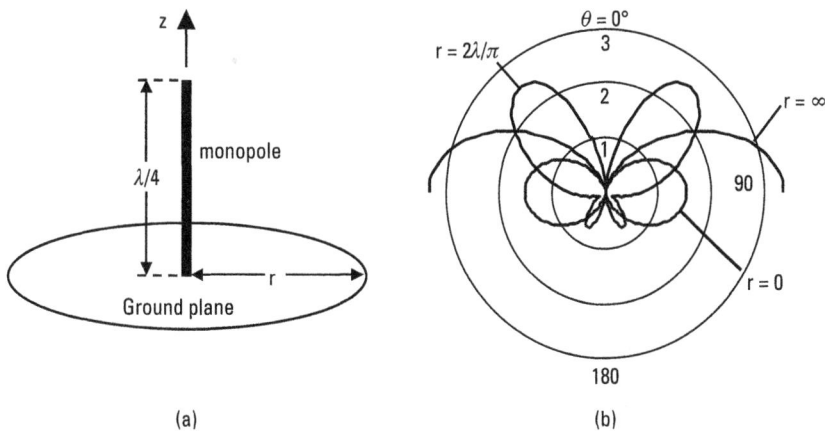

Figure 6.18 (a) Thin, quarter-wave monopole over finite circular ground plane. (b) Radiation pattern for different values of ground plane radius r. (*After:* [12].)

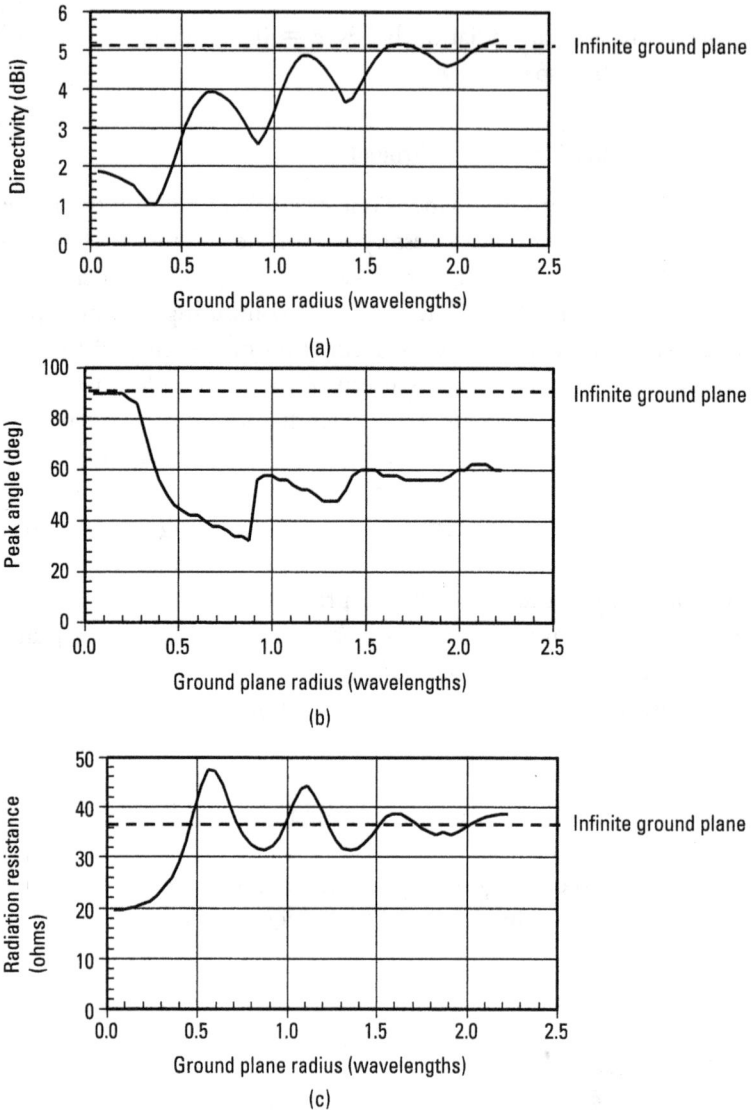

Figure 6.19 Thin, quarter-wave monopole over finite circular ground plane: (a) peak directivity; (b) peak pointing angle; and (c) radiation resistance. (*After:* [12].)

patterns for a monopole slightly less than a quarter-wavelength in height, with a cone that has a length *H* slightly more than a half-wavelength. As the cone angle ψ increases, the angle at which the peak response occurs approaches 0°.

One of the reasons monopole antennas are used with microwave communications equipment is that they can be stowed within the electronics housing and extended when needed. Figure 6.21 shows a monopole protruding through a metal housing. A 50-ohm coaxial transmission line feeds the antenna. The

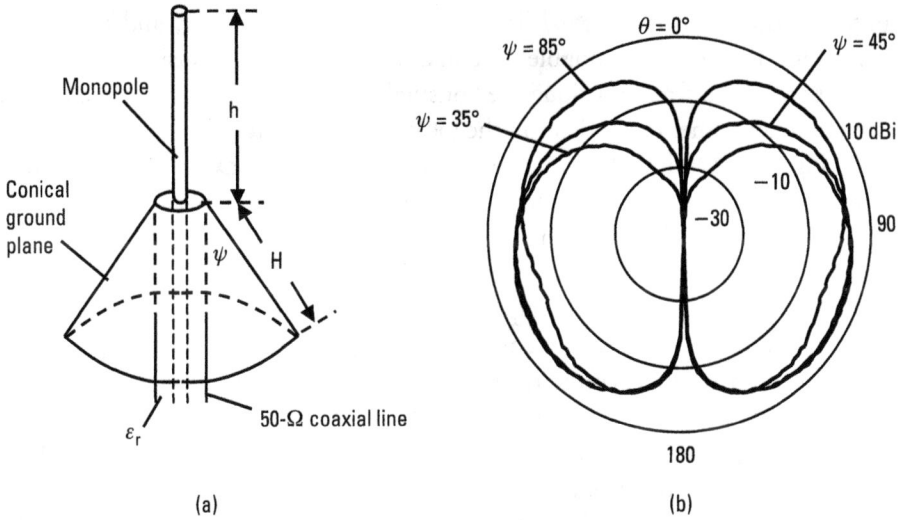

Figure 6.20 (a) Monopole over finite conical ground plane. (b) E-plane pattern versus cone angle at 1.5 GHz. Coaxial line dimensions: ID = 0.1 inch (0.25 cm), OD = 0.377 inch (1.96 cm), ε_r = 2.53, h = 1.85 inches (4.7 cm), H = 4 inches (10.16 cm).

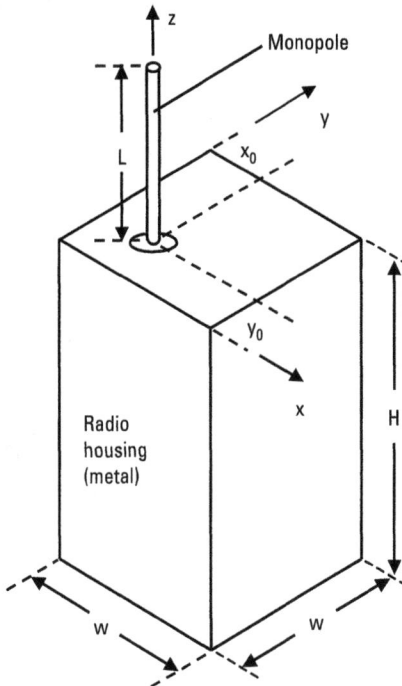

Figure 6.21 The center conductor of a 50-ohm coaxial line forms a monopole antenna above a metal housing. The coaxial outer conductor is grounded to the housing.

outer conductor of the coaxial line is attached to the housing, and the center conductor forms the monopole antenna. The top of the housing images the monopole, and diffraction along the housing's upper edge perturbs the radiation pattern in a manner dependent on the location of the antenna.

Figure 6.22 plots the input match to the 50-ohm coaxial line for two different housing sizes. The monopole is well matched in a narrow bandwidth around 1.5 GHz. Changing the housing width w from 2 inches (5.1 cm) to 3 inches (7.6 cm) shifts the band up 60 MHz, about 4%.

Figure 6.23 plots the principal plane patterns of the monopole. When the housing width is 2 inches (5.1 cm), the antenna is centered in the top surface, the H-plane pattern is symmetric and the E-plane pattern shows deep nulls on axis. When the housing width is 3 inches (7.6 cm), the monopole is off-center, the H-plane pattern is asymmetric, and the E-plane pattern nulls are filled in. Clearly, the size of the ground plane and placement of the coaxial feed contribute significantly to the performance of the antenna installed in the housing. In general, great care must be exercised in the design of housings for systems that rely on low gain antennas, since unanticipated ground plane effects can perturb an antenna's pattern and input impedance in undesirable ways.

The monopole antenna radiation pattern exhibits a null on axis. In applications that require a beam peak orthogonal to the ground plane, a horizontal dipole may be used as shown in Figure 6.24(a). Figure 6.24(b) plots the radiation pattern of the dipole without the ground plane. Figure 6.24(c) plots the pattern with an infinite ground plane at various distances beneath the dipole. When the dipole is very close to the ground plane, the oppositely polarized electric field of the image tends to cancel that of the dipole. At one-quarter wavelength spacing, the dipole and image fields will cancel on broadside, causing a

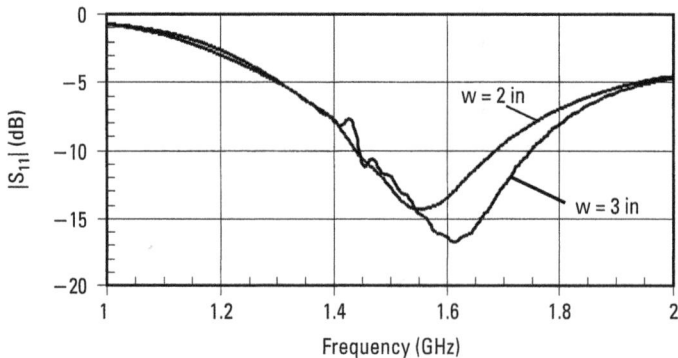

Figure 6.22 Monopole antenna on a metal housing. Input match versus box width. $H = 6$ inches (15.2 cm), $L = 1.85$ inches (4.7 cm), $x_0 = y_0 = 1$ inch (2.5 cm). Monopole diameter: 0.1 inch (0.25 cm).

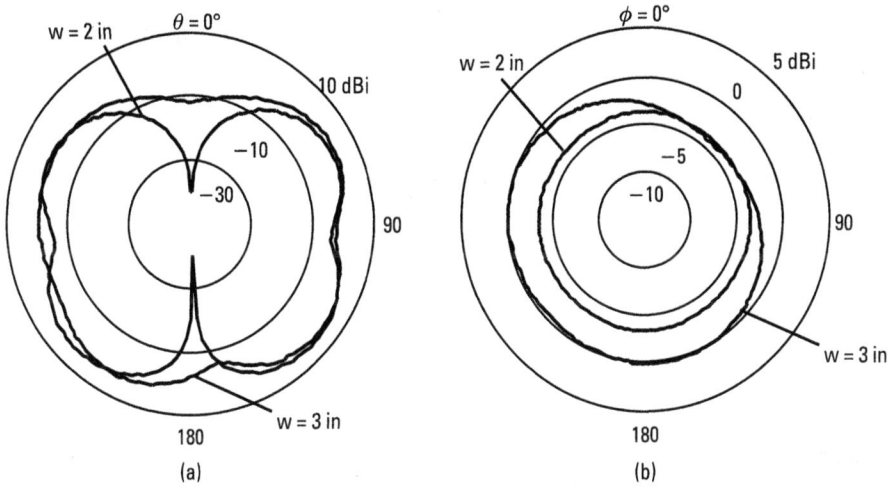

Figure 6.23 Monopole antenna on a metal housing. Principal plane radiation patterns at 1.5 GHz versus box width: (a) E-plane; and (b) H-plane. $H = 6$ inches (15.2 cm), $L = 1.85$ inches (4.7 cm), $x_0 = y_0 = 1$ inch (2.5 cm). Monopole diameter: 0.1 inch (0.25 cm).

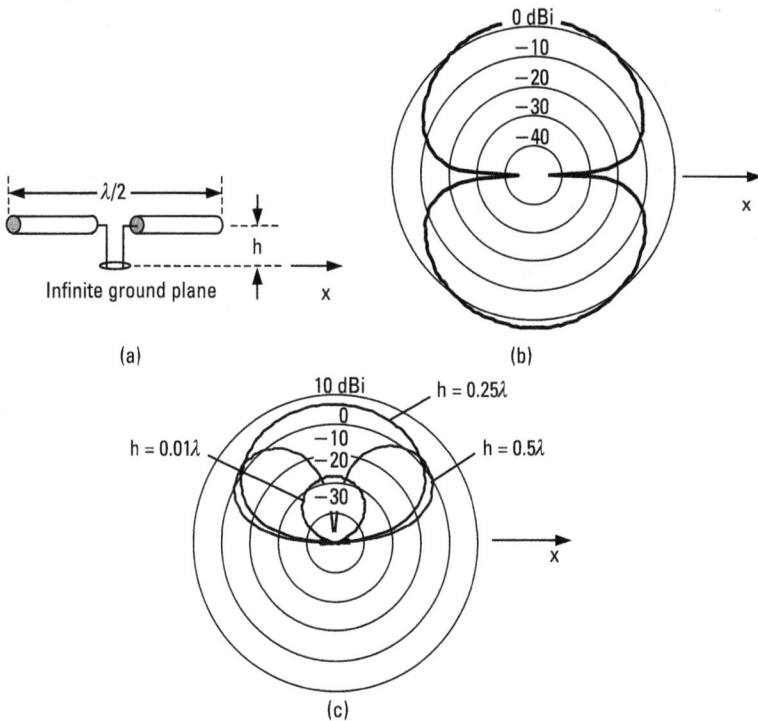

Figure 6.24 (a) Half-wavelength dipole. H-plane radiation pattern of the dipole (b) in free space, and (c) at different heights above an infinite ground plane. Plots normalized to isotropic.

null to appear. For half-wavelength spacing, the fields add constructively to pro-
duce increased gain at broadside.

6.4 Aperture Antennas over Ground Planes

In this section we discuss the performance of a few common aperture antennas
in the presence of ground planes.

6.4.1 Fundamentals

In general, the formation of an opening in a metal structure such as a waveguide
makes an aperture antenna. The simplest aperture radiator is a slot in a ground
plane, which we discussed briefly in Chapter 2. Interruption of current flow is
the basic physical mechanism whereby aperture antennas radiate. When a slot
interrupts current flowing in a ground plane, charge accumulates at its edges and
a radiating electric field forms across the aperture. The slot will radiate equally
well above and below the ground plane. If the slot is half a wavelength long, the
radiation pattern shape is that of a half-wave dipole, but the slot electric field is
polarized across the width of the slot rather than its length. Slots are frequently
cut into waveguide walls, which confine the radiation below the ground plane to
inside the waveguide.

Figure 6.25(a) shows a waveguide horn antenna. Current flows down the
horn from its input until it reaches the horn aperture where it excites an electric
field. To compute the radiation pattern, we assume that the aperture field
amplitude takes the form of a TE_{10} rectangular waveguide mode but with a qua-
dratic phase variation resulting from the varying distance between the horn
input and points on the aperture plane (x-y plane). The radiated electric field in
the E-plane pattern takes the form [13]

$$|E_E(\theta)| = \frac{1+\cos\theta}{2}\left\{\frac{\left[C(r_+)-C(r_-)\right]^2+\left[S(r_+)-S(r_-)\right]^2}{4\left[C^2(2\sqrt{s})+S^2(2\sqrt{s})\right]^2}\right\}^{1/2} \quad (6.8)$$

where

$$r_\pm = 2\sqrt{s}\left(\pm 1 - \frac{B}{4s\lambda}\sin\theta\right) \quad (6.9a)$$

$$s = \frac{B^2}{8\lambda R_2} \quad (6.9b)$$

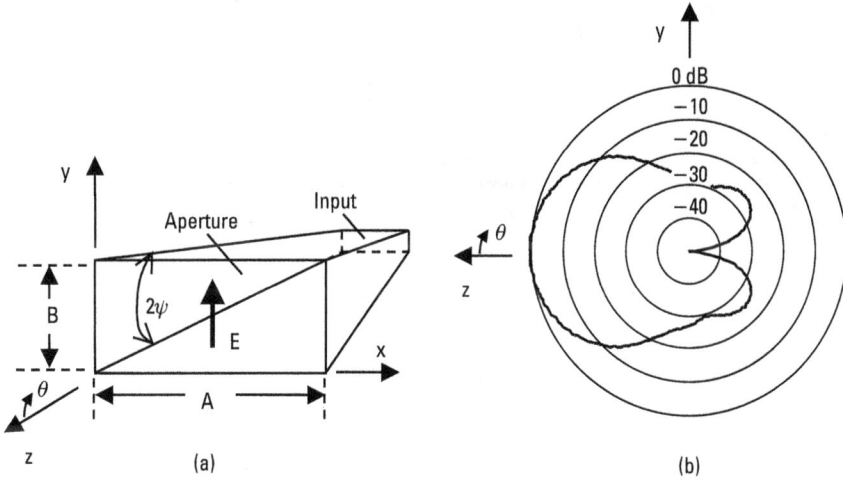

Figure 6.25 (a) Pyramidal horn. (b) E-plane pattern for $B = 1\lambda$, flare angle $\psi = 30°$.

$$R_2 = \frac{B}{2 \tan \psi} \qquad (6.9c)$$

$C(x)$ and $S(x)$ are the Fresnel integrals given by [13]

$$C(x) = \int_0^x \cos\left(\frac{\pi \tau^2}{2}\right) d\tau \qquad (6.10a)$$

$$S(x) = \int_0^x \sin\left(\frac{\pi \tau^2}{2}\right) d\tau \qquad (6.10b)$$

Figure 6.25(b) plots the E-plane radiation pattern of a horn for which $B = 1\lambda$, and $\psi = 30°$. As the figure shows, the aperture radiation is primarily in the forward direction, so a horn antenna is likely to be more directive than a slot or a dipole.

6.4.2 Horn Antennas over Ground

If we place an infinite ground plane beneath and parallel to a horn, as in Figure 6.26(a), the horn field and the image field are normal to the ground plane, and thus they will add like two vertical polarized sources. Figure 6.26(b) plots the radiated far-field pattern of a typical horn with and without the ground plane. The ground plane narrows the beamwidth significantly, and increases the peak gain by a factor of four.

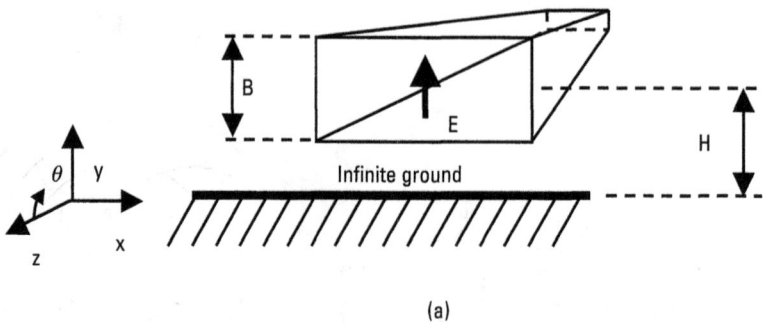

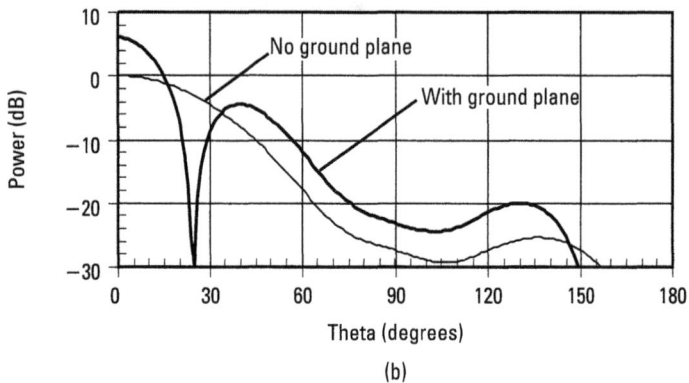

Figure 6.26 (a) Pyramidal horn parallel to an infinite ground plane. (b) E-plane pattern with and without ground plane. $B = 1\lambda$, flare angle $\psi = 30°$. $H = 0.6\lambda$.

A horn antenna also may be placed in front and perpendicular to a ground plane as shown in Figure 6.27(a). The ground plane might be the outer surface of a transceiver electronics enclosure (like the housing in Figure 6.21) to which the horn is attached. Because the horn radiates much less energy to its rear, the ground plane interaction is much less severe then in the parallel case. Even so, for broad beamwidth horns that are used as base station antennas, reflections off a ground plane can cause significant ripple in the radiation pattern as shown in Figure 6.27(b). Without the ground plane, this half wavelength wide aperture has a smooth pattern with an E-plane half-power beamwidth of 86°. With the ground plane present, the beam has significant ripple away from the peak of the beam.

The main beam of larger horn apertures is more directive, so less energy radiates back towards the ground plane. As Figure 6.27(c) shows, the amplitude of the ripple decreases. Another way to reduce ground plane induced ripple is to cover the ground plane with RF absorbing material matched to free-space. This material will absorb the back radiation of the horn rather than reflect it, and in effect, remove the ground plane from the environment.

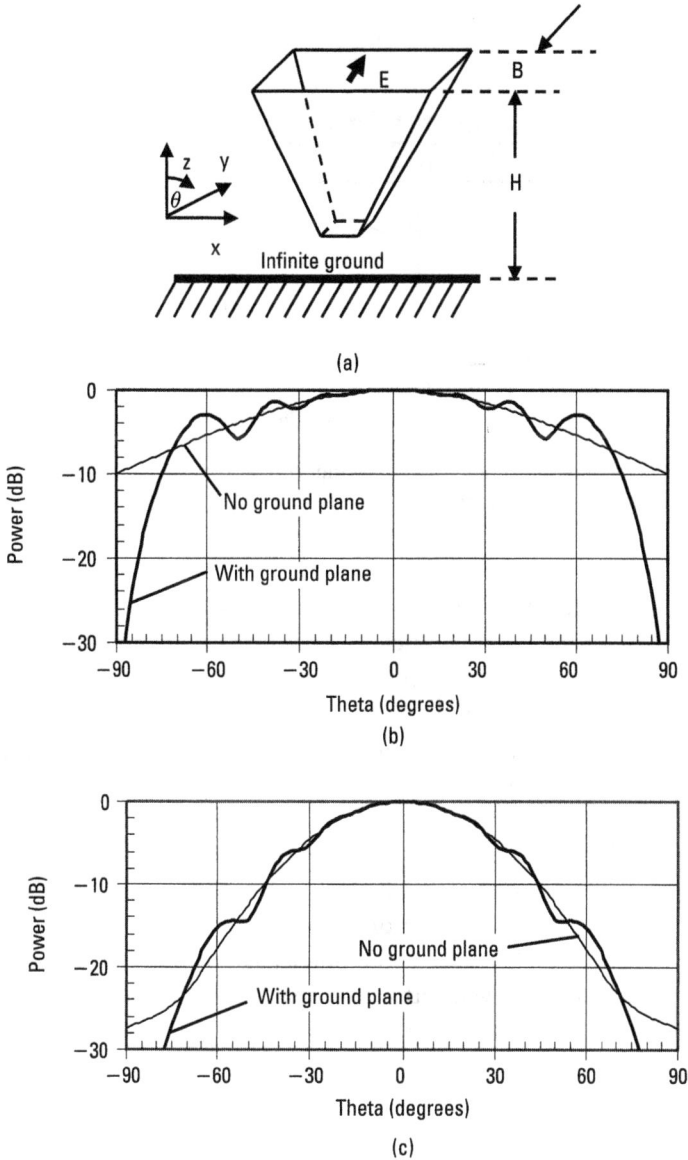

Figure 6.27 (a) Pyramidal horn perpendicular to an infinite ground plane. (b) E-plane pattern with and without ground plane. $B = 0.5\,\lambda$, flare angle $\psi = 30°$, $H = 2\,\lambda$. (c) $B = 1\,\lambda$.

6.4.3   Horn Antennas and Edge Diffraction

Current flowing on the inner surface of a horn like that shown in Figure 6.28 excites the primary aperture field. The radiated field to the rear of the horn is a diffracted field whose source is the current flowing from the mouth of the horn across the outer edge of the aperture wall (see Figure 6.28). The path this

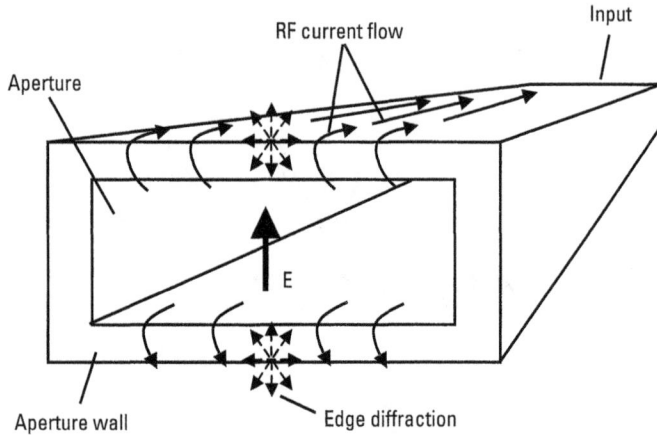

Figure 6.28 Waveguide horn antenna showing primary aperture field and RF current induced edge diffraction.

current follows as it flows back towards ground determines the shape, magnitude, and phase of the diffracted field. Consequently, the thickness of the aperture wall, and even the angles of the intersecting metal surfaces determine how the diffracted field will add to the primary field. The effect of diffraction on the radiation pattern of a horn is most pronounced when the amplitudes of the primary and diffracted fields do not differ greatly, as is the case for broad beam antennas.

Figure 6.29(a) shows a rectangular horn of width B encased in walls of thickness w. Figure 6.29(b) shows how the current flows out of the horn and around its outer wall onto its exterior surface. Diffraction occurs in two places: at the edge of the aperture, and at the outer edge of the wall. For very narrow apertures (less than 0.7 wavelengths), the aperture diffraction can cause significant reflection of the current [14]. For wider apertures, the outer edge

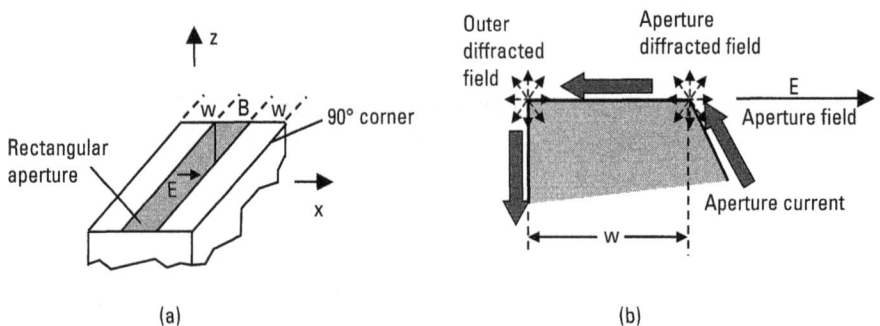

Figure 6.29 (a) Rectangular aperture primary field and diffracted fields. (b) RF current flows around the inner edge of the aperture and is diffracted at outer edge.

diffraction is of more interest, as its interaction with the aperture field distorts the primary radiation pattern. In addition, it is the outer edge that is the source of the field that radiates to the rear of the horn.

For a rectangular aperture embedded in a ground plane [see Figure 6.29(a)], Balanis gives expressions for the diffracted fields at the ground plane edges [15]. We have combined his expressions with the field expressions for a uniformly excited aperture to generate the plots shown in Figure 6.30. The half-power beamwidth of this example horn neglecting diffraction is about 78°. With diffraction included and a wall thickness w of 0.1 wavelengths, the beamwidth decreases to 68°. The beam flattens, and the beamwidth grows to 95° for a wall thickness of half a wavelength. For a wall thickness of one wavelength, the beam shape becomes quite distorted with an inflection point about 30° from the beam peak.

It follows that one can use diffraction of edge current flowing to ground to shape the beamwidth of an aperture antenna. However, often the wall thickness is chosen to meet mechanical requirements, so it cannot be optimized for beam shaping. When the wall thickness exceeds the desired beam shaping thickness, we can use a choke slot like that shown in Figure 6.31. The choke slot is a shorted, quarter-wave, parallel-plate transmission line. At the aperture plane it transforms to an open circuit, effectively stopping the flow of current at the edge nearest the horn aperture. In this way, the choke slot isolates the aperture from any outer structure, and the inner edge of the choke slot can be a precisely placed diffracting edge. Since a choke slot provides a mechanism to control current flow on a ground plane, it may be placed along the edges of an electrically small ground plane to reduce the amplitude of diffracted fields behind the ground plane.

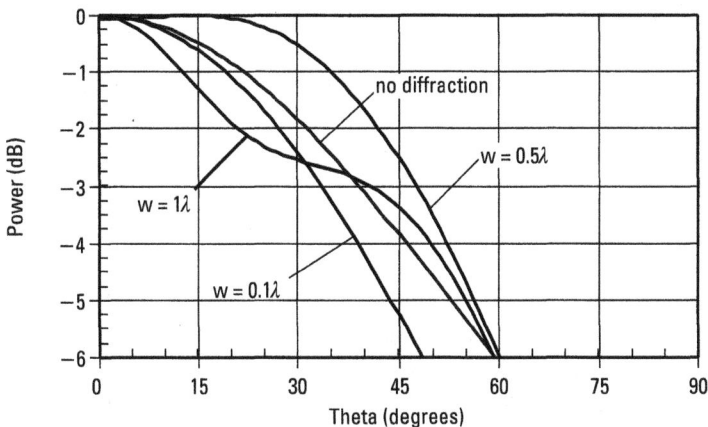

Figure 6.30 E-plane radiation pattern for a uniformly excited aperture versus wall thickness *w*. Aperture width: 0.7λ.

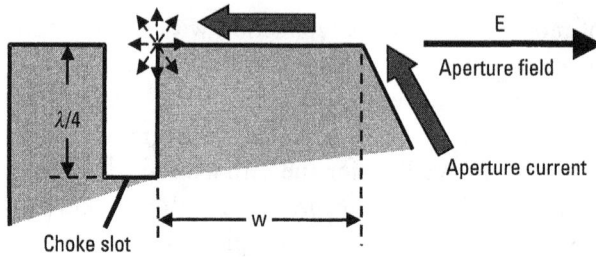

Figure 6.31 A quarter-wave choke slot restricts current flow.

The choke slot is a perfect open circuit only at one frequency; away from that frequency, some current will flow across it. To achieve more bandwidth, one can use two or more choke slots with different dimensions, as shown in Figure 6.32(a). In this case, three choke slots are used to shape the radiation pattern of an antenna into a flat-topped beam that falls steeply on both sides [see Figure 6.32(b)]. This radiation pattern holds up over about a 6% bandwidth [14]. Figure 6.33 shows other conductor structures that modify the flow of aperture current and shape the radiated field of horns [16, 17].

6.4.4   Patch Antennas

The microstrip patch antenna, shown in Figure 6.34(a), is popular because of its very low profile, light weight, and printed circuit fabrication. In the figure, a microstrip transmission line excites the patch. Alternately, a coaxial line can excite the patch orthogonally from below, with the coaxial ground terminating on the microstrip ground plane and the center conductor passing through the substrate to the patch. The patch antenna is an aperture antenna in that it radiates from its ends like a leaky transmission line cavity as shown in Figure 6.34(b). Like a cavity the patch antenna's bandwidth is narrow.

The ground plane's role is fundamental in the operation of the patch. The ground and signal currents at the ends of the patch are the primary sources of the radiated field. The distance between the ground plane and patch sets the height of the cavity and determines the patch's radiation efficiency. In contrast to the microstrip feed line, which requires a thin substrate to suppress radiation, the patch antenna's bandwidth increases with increasing substrate thickness. To resolve this incompatibility between patch and feed requirements, Pozar proposed the aperture-coupled patch antenna shown in Figure 6.35 [18]. Two dielectric layers separated by a common ground plane form the structure, and the patch and microstrip feed are placed on opposite sides of the ground plane. The ground plane has a slot, which couples the feed with the patch. The slot in the ground plane interrupts the microstrip ground current, radiates and excites the patch. A key advantage of this structure is that it allows the thickness H of

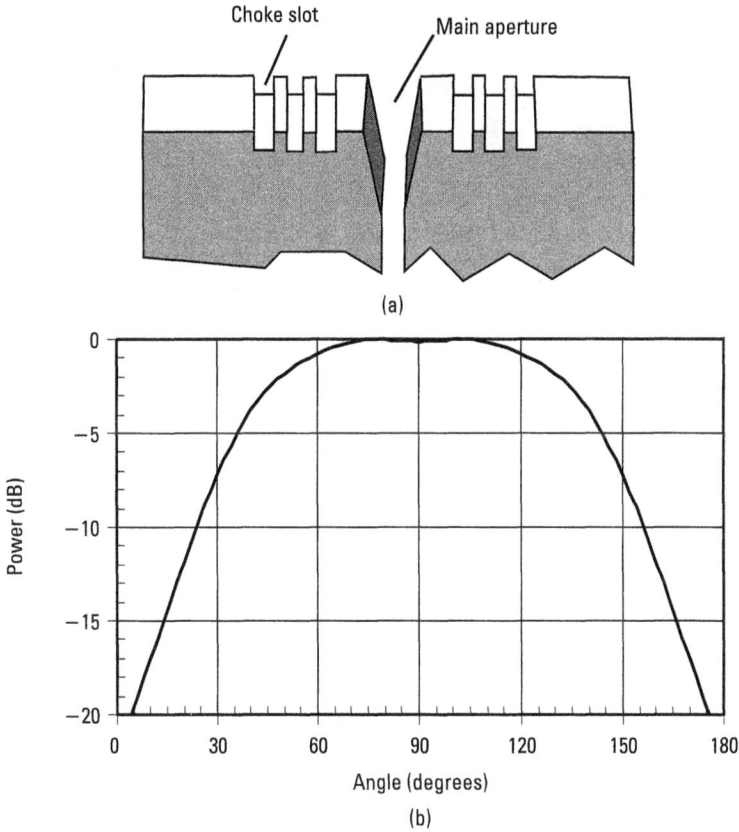

Figure 6.32 (a) Three choke slots used to shape aperture radiation. (b) Flat-topped beam with steep skirts. (*After:* [14].)

the patch substrate to be made large while the thickness h of the microstrip substrate is made thin.

Patch antennas are sometimes placed on very small structures for which the patch's ground plane necessarily must be finite in extent, as shown in Figure 6.36(a). This patch is fed in its center, presumably by the inner conductor of a coaxial transmission line. The patch antenna's radiation pattern is distorted by the diffracting edges of the finite ground plane [19]. As with the horn antenna, the E-plane pattern is affected most by diffraction from currents at the edge of the ground plane. Figure 6.36(b) plots the E-plane pattern for ground planes with different widths L. The radiation pattern is smooth when the ground plane is infinite, but there is significant ripple for finite ground planes.

Namiki et al. have conceived a simple method for eliminating much of the ripple [20]. They modify the diffracting edges of the ground plane by removing rectangular sections of metal as shown in Figure 6.37(a), which causes the diffracting fields to cancel. Figure 6.37(b) compares the measured E-plane patterns

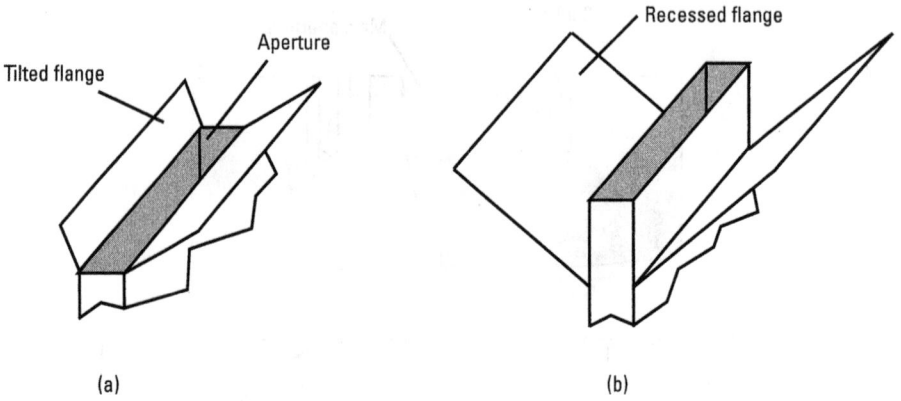

Figure 6.33 Aperture flange structures for shaping aperture radiation: (a) tilted flange; and (b) corner reflector.

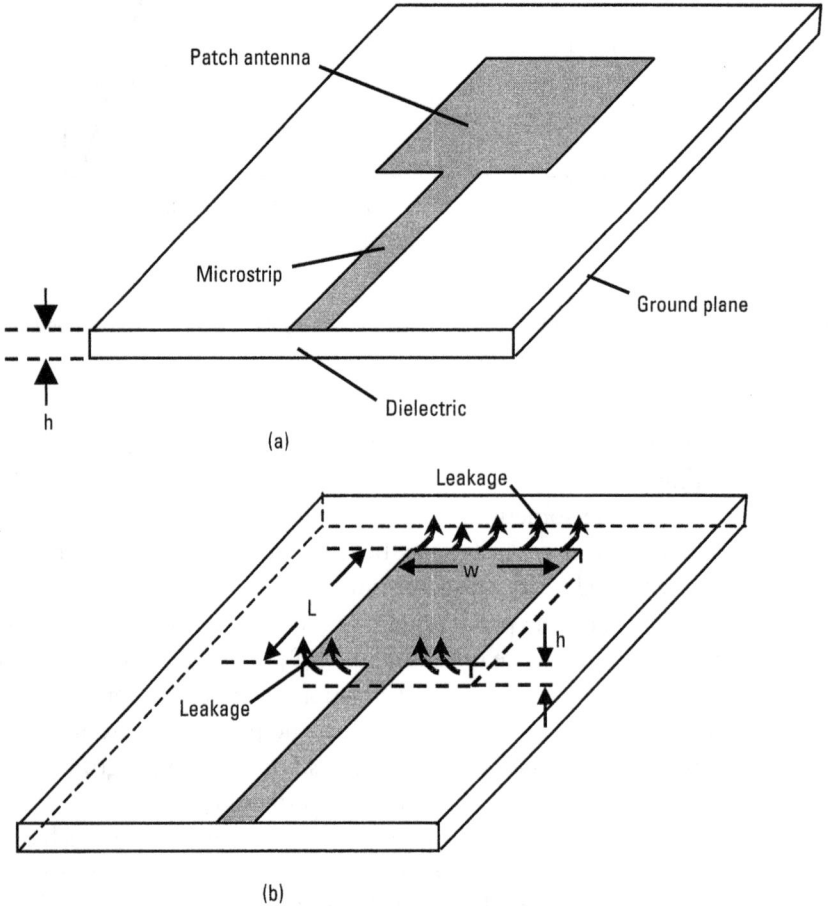

Figure 6.34 (a) Microstrip patch antenna, and (b) leaky cavity radiation model.

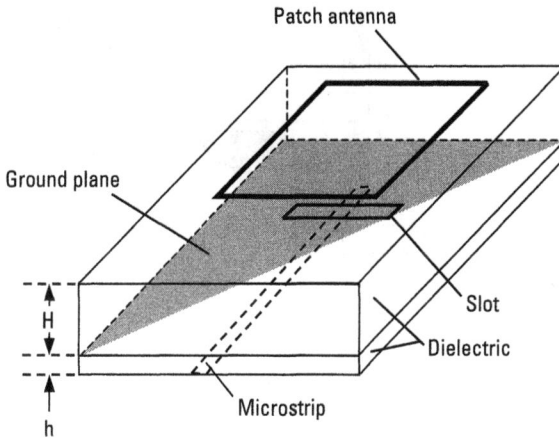

Figure 6.35 A slot coupled patch radiator. Spacing to ground plane is made large for radiator and small for microstrip.

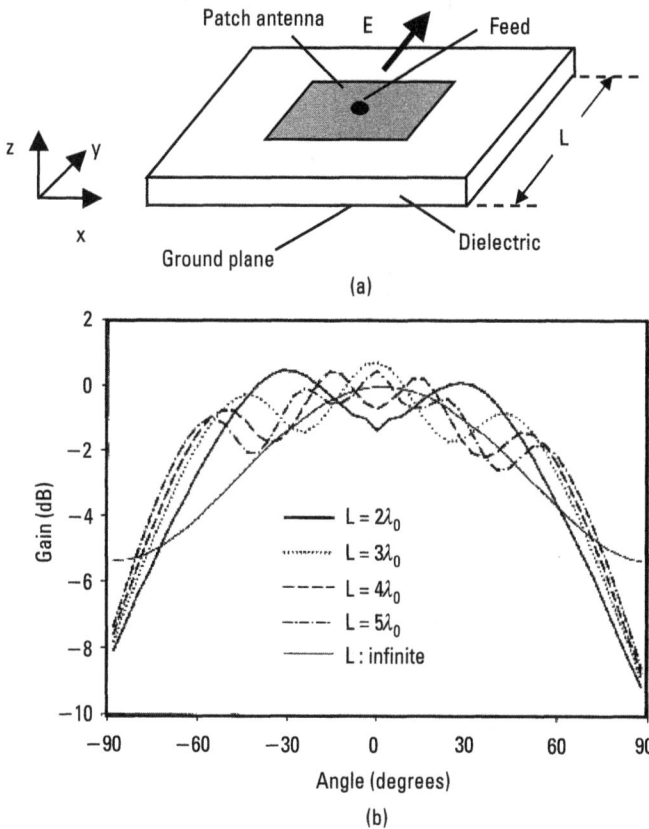

Figure 6.36 (a) Center-fed patch antenna on a finite ground plane. (b) E-plane pattern for different ground plane lengths. (*From:* [20]. © 2003 IEEE. Reprinted with permission.)

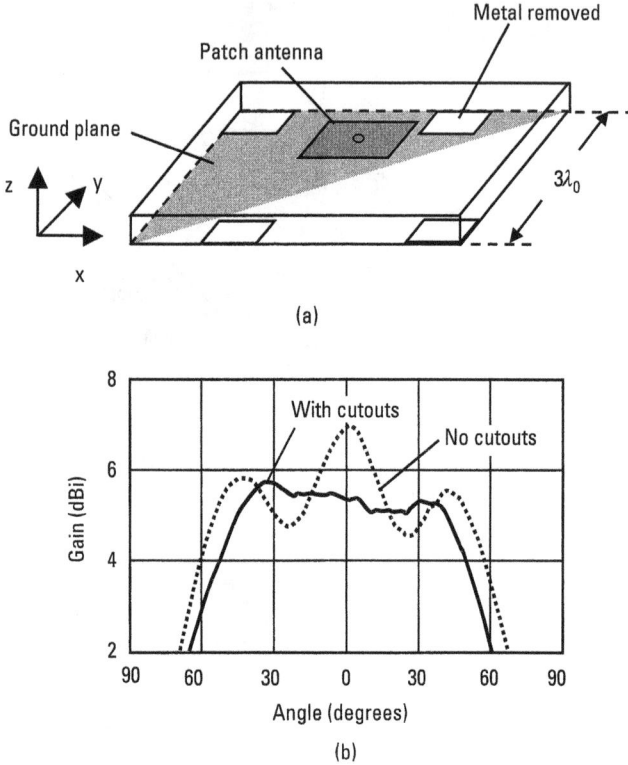

Figure 6.37 (a) Center-fed patch antenna on a finite ground plane with cutouts to cancel diffraction effects in the E-plane. (b) Measured E-plane pattern with and without ground plane cutouts. (*From:* [20]. © 2003 IEEE. Reprinted with permission.)

of a patch on a square ground plane, three wavelengths on a side with and without the edge cutouts. There is a significant reduction in ripple over a bandwidth of about 5%. The ripple in the radiation patterns of other angular pattern cuts perpendicular to the x-y plane is improved also, although less dramatically.

Another antenna, the planar inverted-F antenna (PIFA) shown in Figure 6.38, is much like the patch in construction. It has a radiation pattern that resembles a dipole's, with nulls in the plane of the feeding coaxial probe. This antenna is used in portable wireless devices because it can be made very small. Like a patch antenna, it requires a ground plane to operate, and a coaxial transmission line feeds it. Unlike the patch, the PIFA is shorted to the ground plane on one of its edges.

As with the patch antenna, the PIFA's behavior is strongly influenced by the size of its ground plane. Huynh and Stutzman performed a comprehensive analytical and experimental evaluation of the performance of a PIFA as a function of its ground plane size and shape [21]. They found that the resonant frequency—the frequency at which the PIFA can be matched to the coaxial

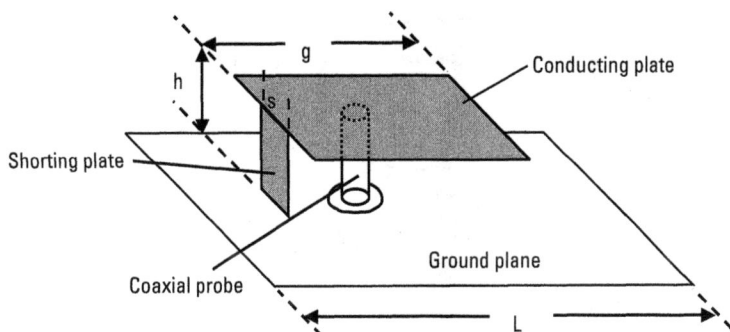

Figure 6.38 Probe-fed PIFA on a finite, square ground plane.

transmission line—has only a slight dependence on ground plane size as long as the dimension L is greater than one quarter of a wavelength. As befits a resonant structure, the PIFA is a narrow bandwidth device, with bandwidth increasing from about 4% at $L = 0.25\lambda$ to 8% to 10% for $L > 0.8\lambda$. A bandwidth of 8% is common for wireless applications. The ground plane edge diffraction introduces high cross polarization, which is desirable for randomly positioned hand-held devices. The PIFA's gain is strongly dependent on ground plane size. It is lowest (about 1 dBi) when $L < 0.4\lambda$, in which case the ground plane and upper conducting plate are about the same size. The gain increases to about 4.5 dBi at $L = 0.9\lambda$, and then oscillates about the infinite ground plane value of about 4.75 dBi as the ground plane size is increased further. A value of 3 dBi, corresponding to $L = 0.5\lambda$, is suitable for most hand-held applications. In addition, the ground plane size significantly influences the PIFA's radiation patterns. Huynh and Stutzman found that the PIFA performed optimally (with high gain and bandwidth) when it was placed near one of the ground plane's corners with the shorting plate at the edge.

6.5 Connecting Antennas to Microwave Circuits

The previous examples in this chapter demonstrate that the operation of an antenna is very dependent on the characteristics of ground planes nearby, be they attached to, or separated from, the antenna. All antennas must be connected to some sort of transmitter or receiver circuit to be of use. We know that such circuits have conductors and ground planes of their own. So it should be no surprise that when we combine a microwave printed circuit board with an antenna, it is likely that the circuit board ground plane will perturb the antenna's performance.

Figure 6.39 shows a 2.4-GHz transceiver circuit board comprising a printed dipole antenna and its associated transceiver circuitry. The circuit board

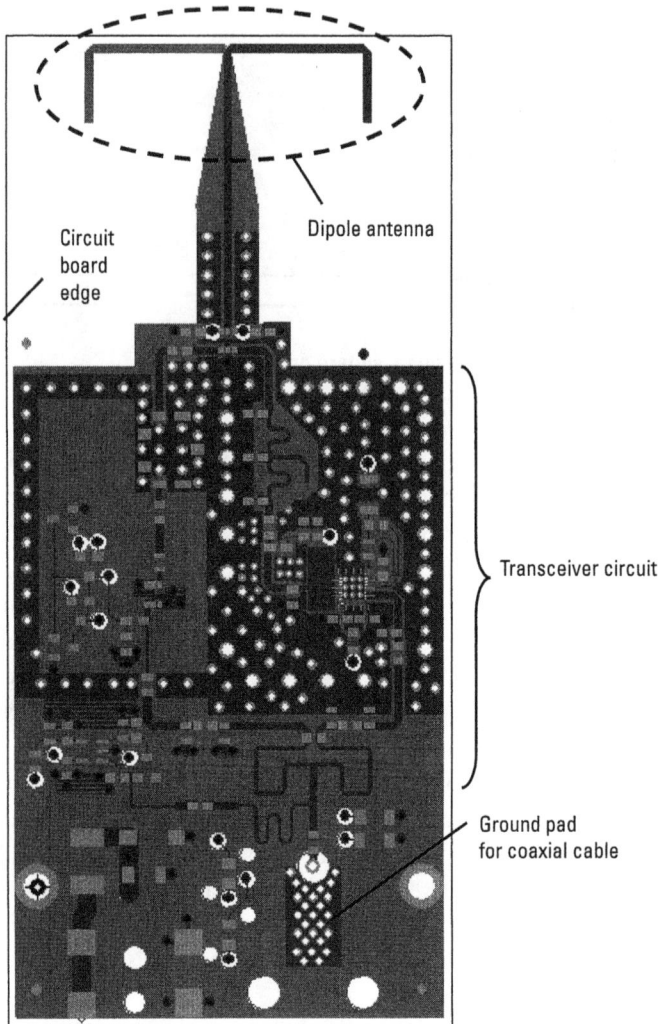

Figure 6.39 Transceiver and dipole antenna on the same circuit board.

is about 4 inches (10 cm) long. At 2.4 GHz, the antenna and transceiver circuit are less than one wavelength apart. Figure 6.40 shows detail of the antenna and the circuit board cross-section. One arm of the dipole is connected to the microstrip signal layer of the board, while the other arm is connected to the microstrip ground plane.

Initially, the antenna in Figure 6.40(a), without the full circuit board ground plane, was optimized. It was then analyzed with the entire transceiver circuit board ground plane attached, as shown in Figure 6.41. Figure 6.42 plots the dipole input match as seen by the feeding 50-ohm microstrip line with and without the full circuit board ground plane. Without the entire circuit board

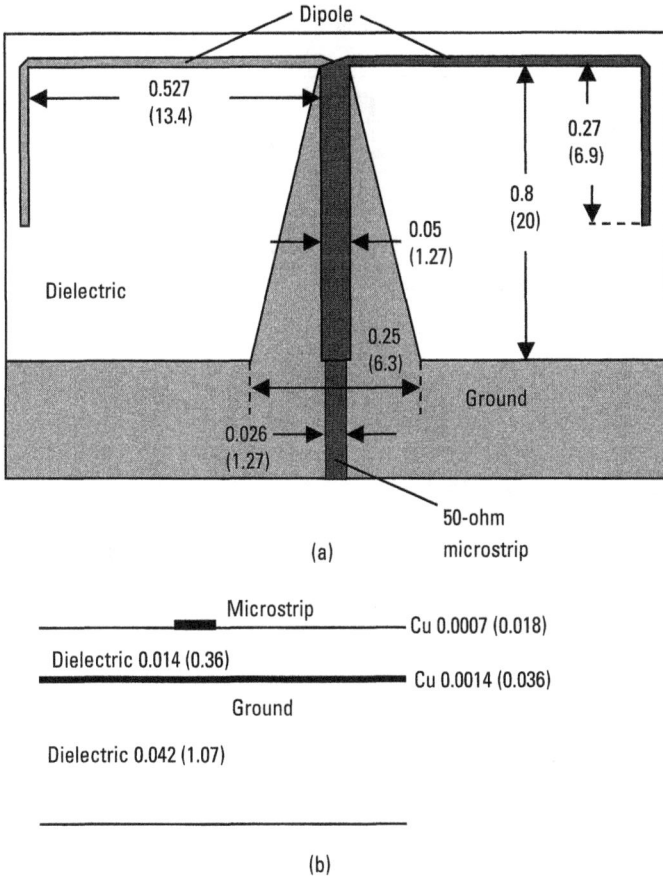

Figure 6.40 Printed dipole antenna: (a) view of microstrip and ground metallization; and (b) circuit board cross-section. Dielectric: FR4 (ε_r = 4.6). Dimensions in inches (millimeters).

present, the input return loss exceeds 12 dB over the desired 2.4- to 2.48-GHz frequency band of operation. With the full circuit board present, the return loss drops significantly, and the antenna must be retuned.

Figure 6.43 plots the E- and H-plane patterns of the dipole with and without the full ground plane attached. In particular, the H-plane pattern, which is almost axially symmetric without the circuit ground plane, exhibits higher directivity above the x-y plane than below it when the full ground is included in the analysis.

The connection between an antenna and the circuit that drives it must have low impedance if the antenna is to work efficiently. Any impedance in the ground plane will add to the antenna's radiation impedance and change the load presented to the feeding transmission line. For example, Figure 6.44 depicts a radio and antenna system found on an automobile. It uses a whip antenna or

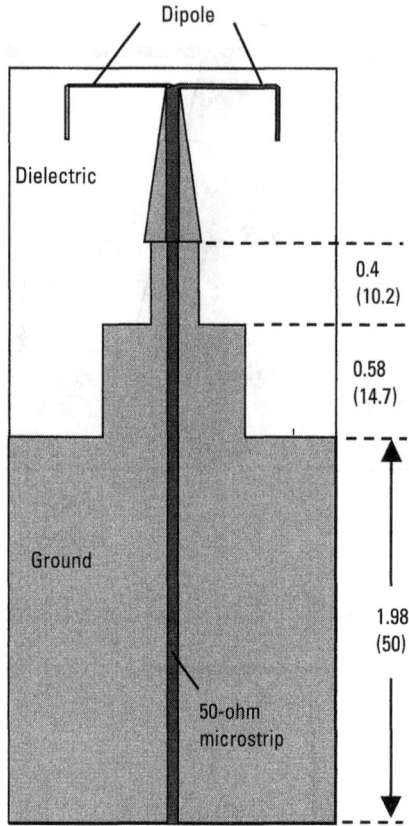

Figure 6.41 Dipole antenna integrated with circuit board ground plane. Dimensions in inches (millimeters).

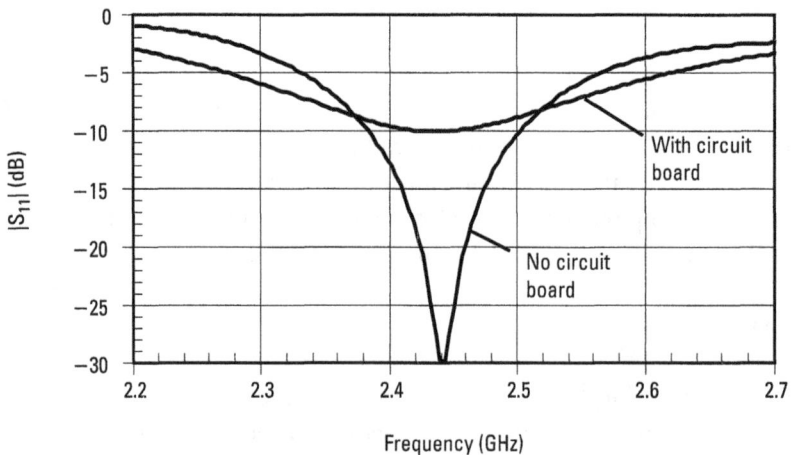

Figure 6.42 Dipole antenna input match ($|S_{11}|$) with and without the circuit board present referenced to 50-ohm microstrip.

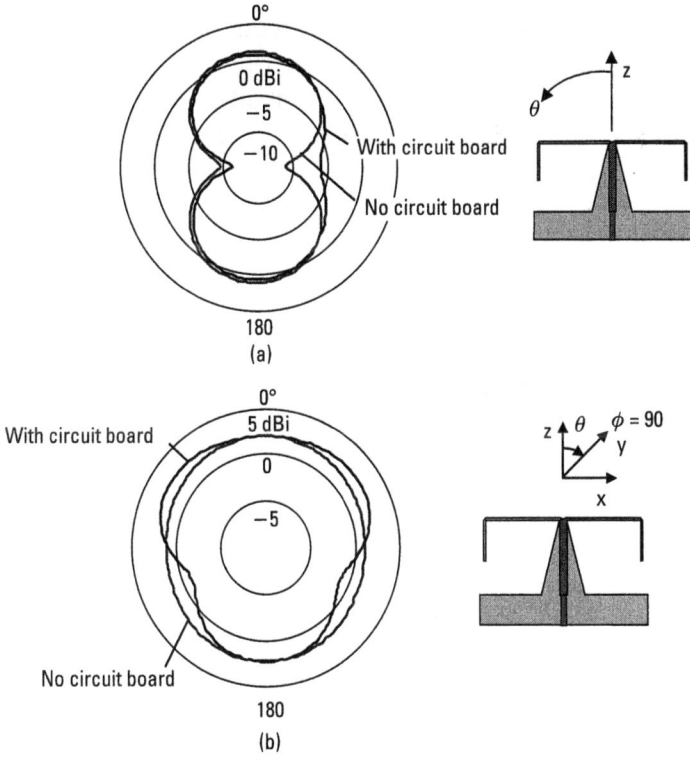

Figure 6.43 Dipole antenna radiation patterns with and without the circuit board present: (a) E-plane; and (b) H-plane. 2.4 GHz.

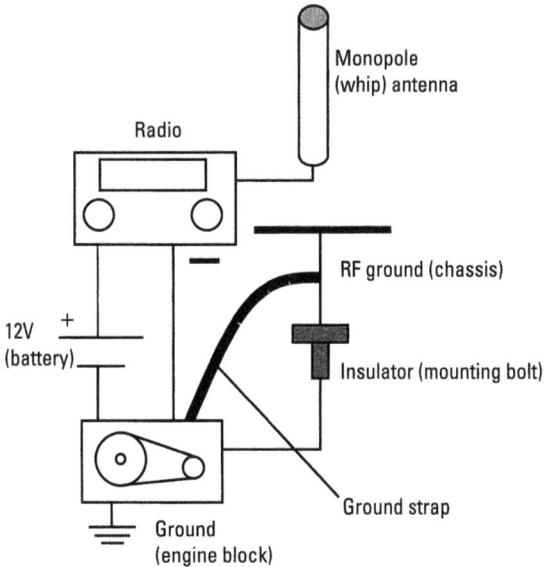

Figure 6.44 Grounding scheme for a mobile radio system. A ground strap compensates for the insulated mounted bolts connecting the RF ground to the engine block.

monopole, and the RF ground plane is the chassis or body of the car. We want the same ground as a reference for the radio, which is powered by the car battery. The car battery's negative terminal and all other electrical grounds typically are connected to the engine block, which is connected to the vehicle chassis through its mounting bolts. Frequently, these bolts are insulated with rubber to suppress vibration. Thus, the antenna ground may be isolated from the radio ground rather than connected via a low impedance path. One can avoid this problem by using a short (low impedance) copper strap to connect the engine block to the chassis [22].

Experienced HAM radio operators are well aware that a good ground connection between their radio and antenna is essential for the maximum receiver sensitivity. Doty and Benson discuss antenna to radio grounding issues in some detail [23, 24].

References

[1]   Kraus, J. D., *Antennas*, 2nd ed., New York: McGraw-Hill, 1988, p. 19.

[2]   Elliott, R. S., *Antenna Theory and Design*, rev. ed., New York: Wiley-Interscience, 2003, p.24.

[3]   Kraus, J. D., *Antennas*, 2nd ed., New York: McGraw-Hill, 1988, pp. 46–47.

[4]   Balanis, C., *Advanced Engineering Electromagnetics*, New York: John Wiley & Sons, 1989, pp. 314–317.

[5]   Kraus, J. D., *Antennas*, 2nd ed., New York: McGraw-Hill, 1988, Section 16-4.

[6]   Stutzman, W. L., *Polarization in Electromagnetic Systems*, Norwood, MA: Artech House, 1993, pp. 183–200.

[7]   Stutzman, W. L., and G. A. Thiele, *Antenna Theory and Design*, New York: John Wiley & Sons, 1981, pp. 90–91.

[8]   Kraus, J. D., *Antennas*, 2nd ed., New York: McGraw-Hill, 1988, p. 179.

[9]   Balanis, C., *Advanced Engineering Electromagnetics*, New York: John Wiley & Sons, 1989, Chapter 13.

[10] Elliott, R. S., *Antenna Theory and Design*, rev. ed., New York: Wiley-Interscience, 2003, p. 63.

[11] Kraus, J. D., *Antennas*, 2nd ed., New York: McGraw-Hill, 1988, Section 16.5.

[12] Weiner, M. M., et al., *Monopole Elements on Circular Ground Planes*, Norwood, MA: Artech House, 1987.

[13] Stutzman, W. L., and G. A. Thiele, *Antenna Theory and Design*, New York: John Wiley & Sons, 1981, pp. 406–408.

[14] Holzman, E. L., "Pillbox Antenna Design for Millimeter-Wave Base Station Applications," *IEEE Antennas & Propagation Magazine*, Vol. 45, No. 1, February 2003, pp. 27–37.

[15]   Balanis, C., *Advanced Engineering Electromagnetics*, New York: John Wiley & Sons, 1989, pp. 844–845.

[16]   Koshy, V. K., K. G. Nair, and G. P. Srivastava, "Analysis of Radiation from a Flanged Aperture Antenna," *IEEE Trans. on Antennas and Propagation*, Vol. AP-18, 1970, pp. 407–411.

[17]   Nair, K. G., and P. A. Matthews, "Metallic Flanges on Sectoral Horns as Corner Reflectors," *IEEE Trans. on Antennas and Propagation*, Vol. AP-25, 1977, pp. 431–435.

[18]   Buck, A. C., and D. M. Pozar, "Aperture-Coupled Microstrip Antenna with a Perpendicular Feed," *Electron Lett.*, Vol. 22, 1986, pp. 125–126.

[19]   Huang, J., "The Finite Ground Plane Effect on the Microstrip Antenna Radiating Patterns," *IEEE Trans. on Antennas and Propagation*, Vol. AP-31, 1983, pp. 649–653.

[20]   Namiki, T., Y. Murayama, and K. Ito, "Improving Radiation-Pattern Distortion of a Patch Antenna Having a Finite Ground Plane," *IEEE Trans. on Antennas and Propagation*, Vol. AP-51, 2003, pp. 478–482.

[21]   Huynh, M.-C., and W. Stutzman, "Ground Plane Effects on Planar Inverted-F Antenna (PIFA) Performance," *IEE Proc.-Microw. Antennas Propag.*, Vol. 150, No. 4, August 2003, pp. 209–213.

[22]   Johnson, B. W., "RF Ground and the Ground Plane," http://www.k2bj.com/Ground.htm.

[23]   Doty, J., "Grounding Is Key to Good Reception," http://www.hard-core-dx.com/nordicdx/antenna/ground/ground2.html.

[24]   Benson, T., "Ground Zero for Your Receiver," http://www.hard-core-dx.com/nordicdx/antenna/ground/ground1.html.

About the Author

Eric Holzman holds a B.S., an M.S., and a Ph.D. in electrical engineering from the University of California at Los Angeles. He joined Northrop Grumman Electronic Systems in Baltimore, Maryland, in 2004 as a consulting engineer in the Advanced RF Product Technology Department. His work involves the design and analysis of active phased arrays and other antennas operating from UHF to millimeter-wave frequencies. From 1999 until 2004, he was a senior microwave engineer at YDI Wireless, where he designed antennas and transceiver circuits for a variety of commercial fixed wireless applications. From 1993 until 1999, he was a principal engineer and manager at Lockheed Martin Government Electronic Systems, where he was involved in the design of advanced, solid-state phased arrays. He began his career designing power oscillators and low noise amplifiers and antennas for Hughes Missile Systems Company.

Dr. Holzman is a Senior Member of the IEEE, a former reviewer for the *IEEE Transactions on Antennas and Propagation*, and a past chairman of the Philadelphia Chapter of the IEEE Antennas/Microwave Societies. He was a member of the organizing committee for the Benjamin Franklin Symposium in 1995, 1996, and 1997. He is the lead author of the book *Solid-State Microwave Power Oscillator Design* (Artech House, 1992), the author of 35 publications, and the holder of seven patents in the microwave field. He was the Lockheed Martin "Engineer of the Year" in 1997 for his work on antennas and transmit/receive modules. He is a former Howard Hughes Fellow, listed in *Who's Who in Young America* (1992), and is a member of Tau Beta Pi and Eta Kappa Nu.

Index

Recent Titles in the Artech House Microwave Library

For further information on these and other Artech House titles, including previously considered out-of-print books now available through our In-Print-Forever® (IPF®) program, contact:

Artech House Publishers

685 Canton Street

Norwood, MA 02062

Phone: 781-769-9750

Fax: 781-769-6334

e-mail: artech@artechhouse.com

Artech House Books

46 Gillingham Street

London SW1V 1AH UK

Phone: +44 (0)20 7596 8750

Fax: +44 (0)20 7630 0166

e-mail: artech-uk@artechhouse.com

Find us on the World Wide Web at: www.artechhouse.com

www.ingramcontent.com/pod-product-compliance
Lightning Source LLC
Chambersburg PA
CBHW031950180326
41458CB00006B/1680